P9-CSB-453

The GREATEST SCIENCE STORIES NEVER TOLD

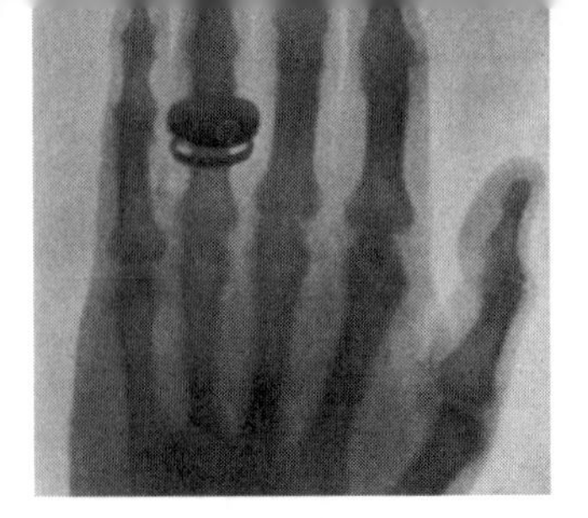
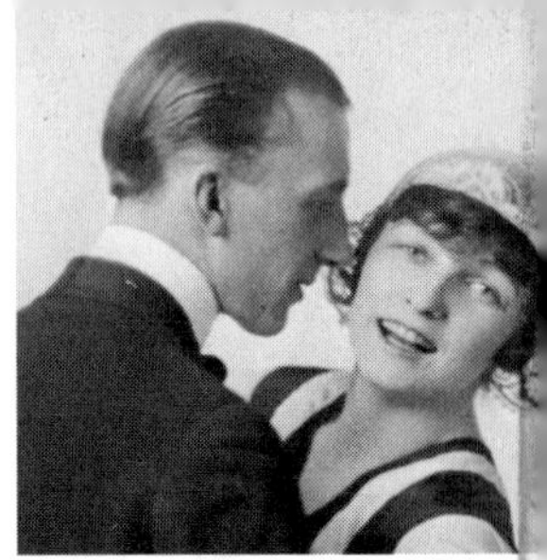

ALSO BY RICK BEYER

The Greatest Stories Never Told
The Greatest War Stories Never Told
The Greatest Presidential Stories Never Told

HARPER
An Imprint of HarperCollins*Publishers*
www.harpercollins.com

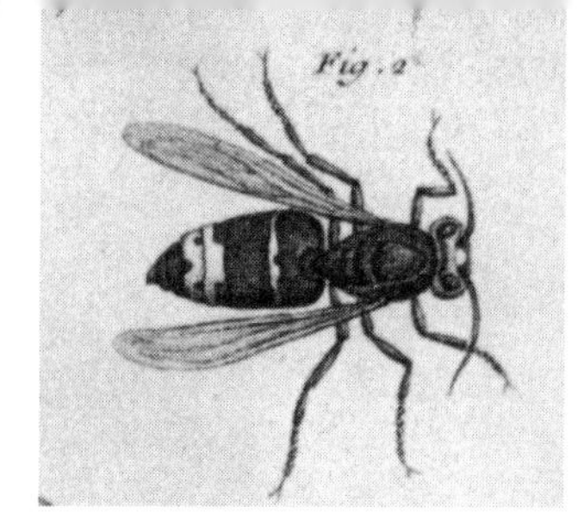

presents . . .

The GREATEST SCIENCE STORIES NEVER TOLD

100 *Tales of Invention and Discovery to Astonish, Bewilder, & Stupefy*

by Rick Beyer

HarperCollins books may be purchased for educational, business, or sales promotional use. For information, please write: Special Markets Department, HarperCollins Publishers, 10 East 53rd Street, New York, NY 10022.

FIRST EDITION

Library of Congress Cataloging-in-Publication Data

Beyer, Rick.
The greatest science stories never told: 100 tales of invention and discovery to astonish, bewilder, & stupefy / Rick Beyer. — 1st ed.
p. cm.
At head of title: History Channel presents—
ISBN 978-0-06-162696-8
1. Science—Miscellanea. I. History Channel (Television network) II. Title. III. Title: History Channel presents—.
Q173.B593 2009
500—dc22 2009032741

09 10 11 12 13 OV/QW 10 9 8 7 6 5 4 3 2 1

In memory of two inspirational scientists who will also be remembered as wonderful human beings: Robert T. Beyer (1920–2008) and Charles Gordon Zubrod (1914–1999).

INTRODUCTION

Thomas Edison did not invent the lightbulb. Alexander Graham Bell did not invent the telephone. The first car was careening around Paris before George Washington was president. The vending machine is nearly two thousand years old. The fax machine was invented by an Italian priest during the Civil War. The pursuit of a death ray led to the invention of radar, which melted another inventor's candy bar, inspiring the invention of the microwave. The inventor of the lie detector created Wonder Woman. E-mail was originated by somebody trying to avoid his "real work." And the idea behind modern television was dreamed up by a teenage boy plowing a potato field.

Want more? Good! You've come to the right place.

This is the fourth in my series of Greatest Stories Never Told books, published in conjunction with The History Channel®. It is full of adventure stories, happy accidents, lifelong obsessions, and blinding flashes of brilliance. True tales from two thousand years of invention and discovery, populated with bigger-than-life characters such as Isaac Newton (a secret alchemist and fearsome battler of counterfeiters) and little-known inventors like Albert Parkhouse (the man who gave us the wire coat hanger).

The sudden revelation of a "Eureka!" moment can happen at any time or place. It famously hit Archimedes in the bathtub, as you can read in the book's first story. Inspiration came to one scientist in a dream, to another at midnight on a park bench, to a third while he was riding a trolley car. In each case, the moment led to a Nobel Prize. Serendipity often plays a telling role. Teflon, penicillin, the X-ray, safety glass, and saccharin are just a few of the things that were stumbled upon by accident. A walk in the woods in 1719 led to the development of modern paper, while another nature stroll a few centuries later inspired the invention of Velcro. And astronomers who were trying to eliminate the hiss from their radio telescope accidentally discovered proof of the Big Bang theory.

An important part of science is taking the theoretical and applying it to the real world, by developing new technologies and products. I granted myself the latitude to include stories about a wide range of such endeavors in the pages that follow. The result runs the gamut, from Archimedes to the Zamboni.

A distinguished physicist with a sense of whimsy once said that "nothing was invented for the first time." There are plenty of stories in here about the forgotten inventors who developed such things as radio, the movie camera, and the computer well before the famous names who ended up with the credit. And don't forget the early movie star who invented the turn signal or the scandalous socialite who was the first to patent a brassiere!

Lest you think the annals of science make for dry reading, let me assure you that the opposite is true. There are suicides, mysterious disappearances, a duel to the death, earthquakes, and a surprising amount of golf! Some scenes are so striking they belong in movies: two hundred monks in a giant circle being simultaneously jolted by an electrical shock, or a railroad flatcar filled with trumpet players sounding the same note as the train races through the station. Who knew the pursuit of science could be so amazing?

Some of the stories here I learned from my father, Robert T. Beyer, a scientist and author with a lifelong interest in history. Others I turned up during countless hours of research, experiencing my own eureka epiphanies each time I came across something delicious that I didn't know. My goal is to get readers as excited about these history gems as I am. Just don't get up from the bathtub and run naked through the streets to tell people about them. That's so been done.

The GREATEST SCIENCE STORIES NEVER TOLD

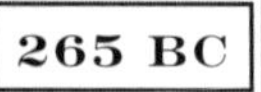

265 BC

BATH TIME

The original eureka moment

It may be the most famous and dramatic moment in the history of bathing—not to mention science: Archimedes leaping from his bath and running down the street naked, shouting *Eureka!*—Greek for "I found it!" To this day, when inspiration descends in a blinding flash, it is referred to as a "Eureka!" moment.

But just what was it that he was so excited about?

Archimedes lived in the Greek city-state of Syracuse, on the island of Sicily. King Hiero of Syracuse ordered a gold crown be made as an offering to the gods. He gave the appropriate amount of gold to a goldsmith, who created a beautiful crown. But then rumors spread that the goldsmith had stolen some of the gold, and substituted cheaper silver in the crown. The question was how to find out for sure—without damaging the crown. The king put Archimedes on the case.

Archimedes was Einstein and Edison combined, the greatest scientist of the ancient world, and also a brilliant inventor. Clearly, the right man for the job.

He was pondering the problem one day at the baths. As he stepped into a full tub, he noticed the water rising. Suddenly, it came to him that he could measure the *volume* of the crown by seeing how much water it displaced. That was the key to unraveling the mystery.

Archimedes determined that the crown was at least partly silver. Life-shortening news for the goldsmith, no doubt, but a long-lived story for the annals of science.

Archimedes went on to develop fundamental principles of buoyancy, which he outlined in a treatise called On Floating Bodies. *It is a pioneering work in the field of hydrostatics.*

Here's the exact solution Archimedes came up with. He weighed the crown. Then he created a lump of gold and a lump of silver that were each the same weight as the crown. He submerged each of them in turn in a container of water to see how much water they displaced. The silver displaced the most. The gold, being less dense, displaced the least. The crown was right in the middle, suggesting it was a mixture of silver and gold. *Eureka!*

Did it really happen? The Eureka! *story isn't mentioned in any of the surviving writings of Archimedes. The earliest known telling of it came more than two hundred years later, in a book by the great Roman architect and engineer Vitruvius, although there may have been earlier versions of the story, which didn't survive. Archimedes may well have tested the crown for King Hiero, but most historians think the part about him running naked through the streets was just a way to dress up the story a bit.*

ΑΡΧΙΜΗΔΟΥΣ

ΑΡΧΙΜΗΔΟΥΣ ΚΥΚΛΟΥ ΜΕΤΡΗΣΙΣ.

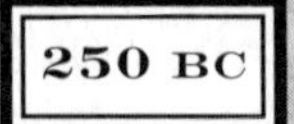

250 BC

PERSIAN POWER

The enigmatic batteries of Baghdad

We don't usually think of people who lived more than two thousand years ago as using electricity. But, in fact, batteries might have been state-of-the-art in ancient Persia.

In 1938, archeologists excavating a tomb outside Baghdad found a small clay jar unlike any previously discovered. It was about five inches tall, with a copper cylinder in the center, and an iron rod inside that. The metal was corroded, and tests showed that the jar had once held some sort of acid, possibly vinegar.

Was it, perhaps, a battery?

It was constructed just like a simple battery, and many scientists believe that's exactly what it was. Tests performed on a replica show that when filled with vinegar, it can generate up to one volt of electricity.

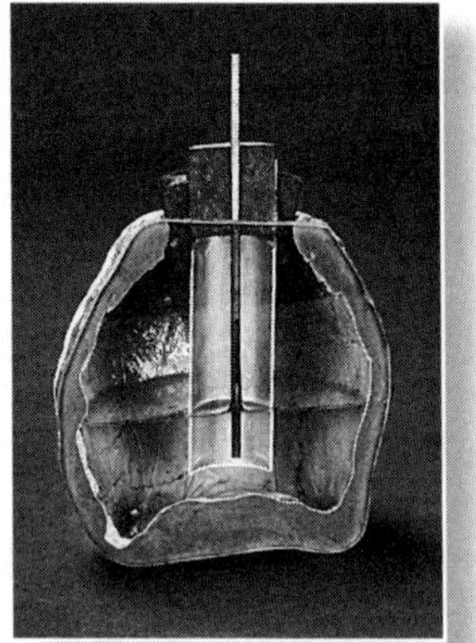

Over the years, at least a dozen similar jars have been found. They are known as the Baghdad batteries. What might they have been used for? One theory is that many batteries could have been hooked up together and used for electroplating jewelry. Some of the batteries were found alongside needles, suggesting that perhaps they were used to provide a little extra punch to ancient acupuncture. Still another idea is that they were employed by priests in the temple to inspire awe in the worshippers.

So perhaps the phrase "battery-powered" isn't as modern as we think.

Some of the jars were found near the palace of the Parthian King, Ctesiphon.

Not everyone thinks the jars were really batteries. Skeptics point out that no wires have ever been discovered with them, and suggest they were some sort of sacred document holders. No documents, however, have ever been discovered inside any of the jars.

62

ANCIENT HERO

The Greek inventor who got there first

How long do you suppose the coin-operated vending machine has been around? Chances are your guess would fall short of the correct answer: nearly two thousand years.

The first one was invented by an extraordinary Greek scientist named Hero, who lived in the Egyptian city of Alexandria. It dispensed water in the temple. Worshippers could put in a five-drachma coin and get a little water to wash their face and hands.

The showier the invention, the more Hero seemed to like it. He made automatic doors for the temple, a mechanical puppet theater, and a solar-

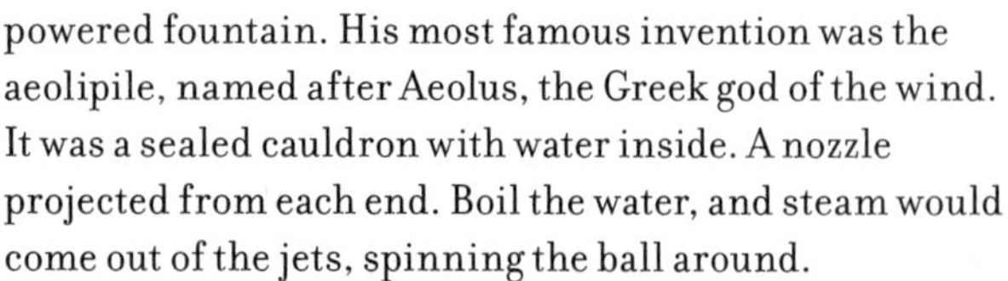

powered fountain. His most famous invention was the aeolipile, named after Aeolus, the Greek god of the wind. It was a sealed cauldron with water inside. A nozzle projected from each end. Boil the water, and steam would come out of the jets, spinning the ball around.

It was the forerunner of the steam engine and the jet engine—but way ahead of its time.

Hero never quite managed to harness its power. He had invented all the necessary parts, but he never figured out how to link them together. Some say he died trying. Had he succeeded, it might have led to a Roman Industrial Age. And all history since might be very different.

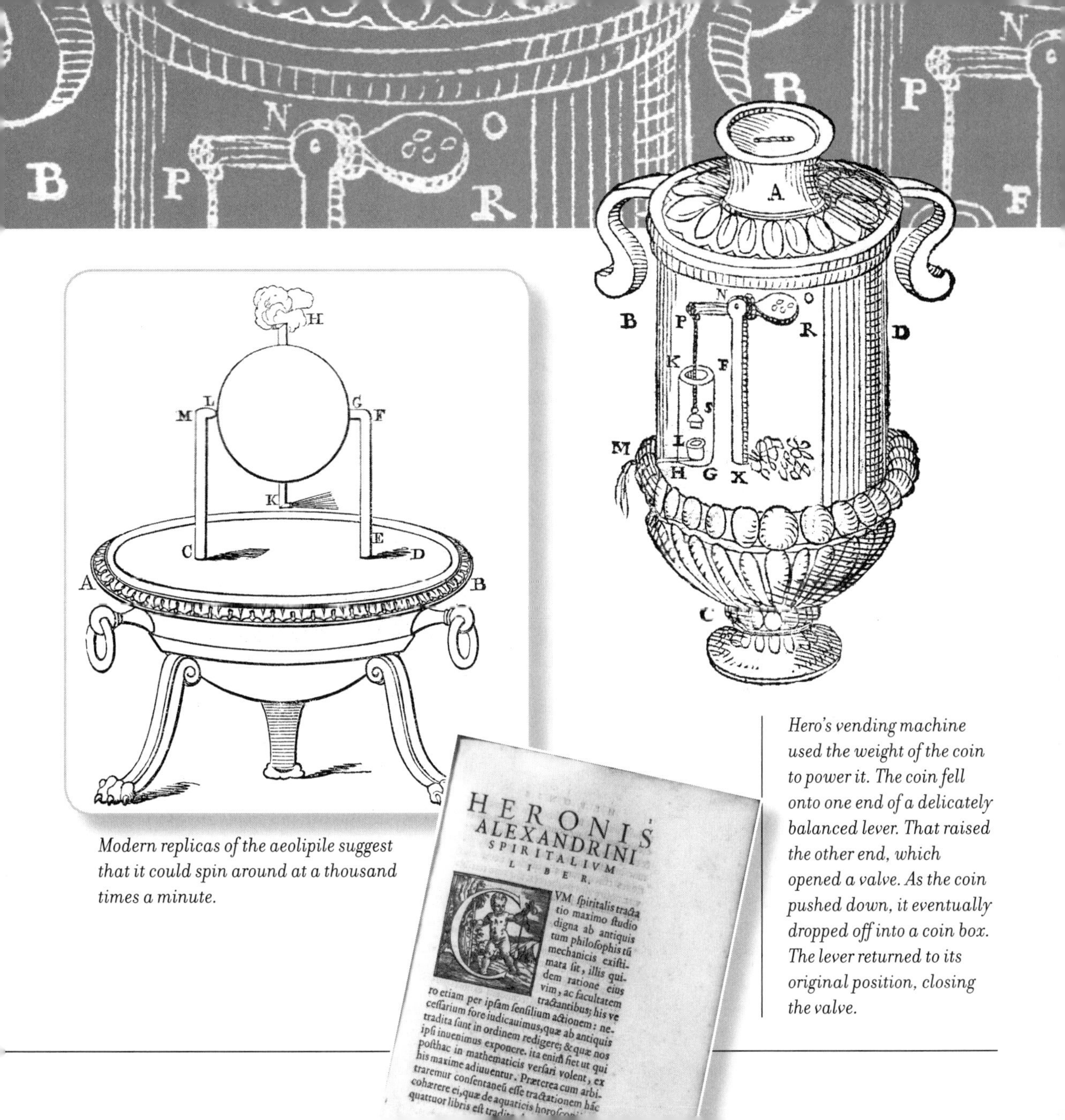

Modern replicas of the aeolipile suggest that it could spin around at a thousand times a minute.

HERONIS
ALEXANDRINI
SPIRITALIVM
LIBER.

CVM ſpiritalis tracta tio maximo ſtudio digna ab antiquis tum philoſophis tũ mechanicis exiſtimata ſit, illis quidem ratione eius vim, ac facultatem tractantibus; his vero etiam per ipſam ſenſilium actionem: neceſſarium fore iudicauimus, quæ ab antiquis tradita ſunt in ordinem redigere; & quæ nos ipſi inuenimus exponere. ita enim fiet ut qui poſthac in mathematicis verſari volent, ex his maxime adiuuentur. Præterea cum arbitraremur conſentaneũ eſſe tractationem hãc cohærere ei, quæ de aquaticis horoſcopiis quattuor libris eſt tradita

Hero's vending machine used the weight of the coin to power it. The coin fell onto one end of a delicately balanced lever. That raised the other end, which opened a valve. As the coin pushed down, it eventually dropped off into a coin box. The lever returned to its original position, closing the valve.

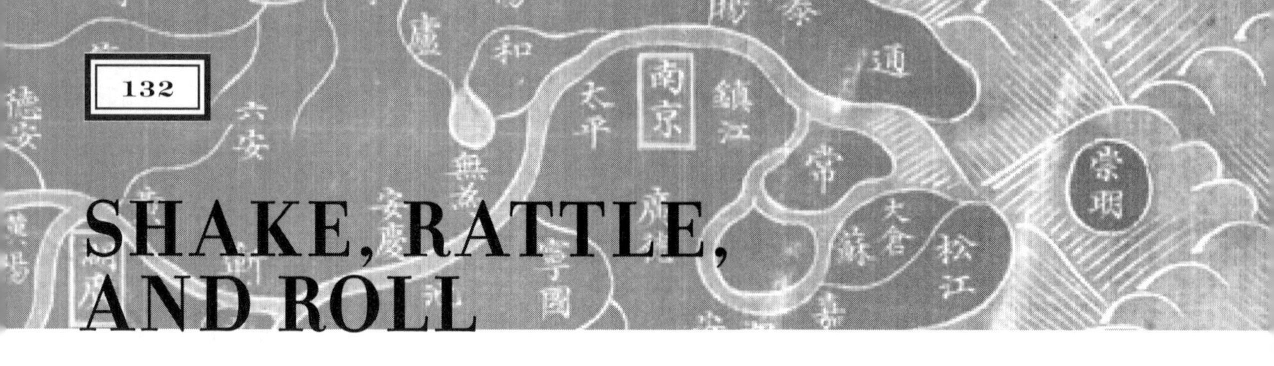

132

SHAKE, RATTLE, AND ROLL

A story of dragons, toads, and earthquakes

A seismometer is a sophisticated device for measuring earthquakes. But instead of being the modern invention you might imagine, the first seismometer was built in China nearly two thousand years ago.

It was built by a man as famous in China as he is unknown in the West: Zhang Heng. He served as chief astronomer of the Eastern Han Dynasty, but that doesn't begin to capture his varied accomplishments. He was also a mathematician, inventor, geographer, artist, poet, and civil servant.

China has long been plagued by ferocious earthquakes. Zhang sought to create something that could detect earthquakes in distant areas, and predict a major shock by detecting tremors that might precede it.

The ingenious device he invented took the form of an ornately decorated copper barrel about eight feet in diameter. Eight dragon-heads protruded in different directions, each holding a copper ball in its jaws. Below every dragon was a toad with an upturned mouth. Inside the barrel was a sensitive mechanism of pendulums and levers. When it felt a tremor, however slight, a ball would drop out of the mouth of a dragon into the mouth of a toad, showing from which direction the quake came.

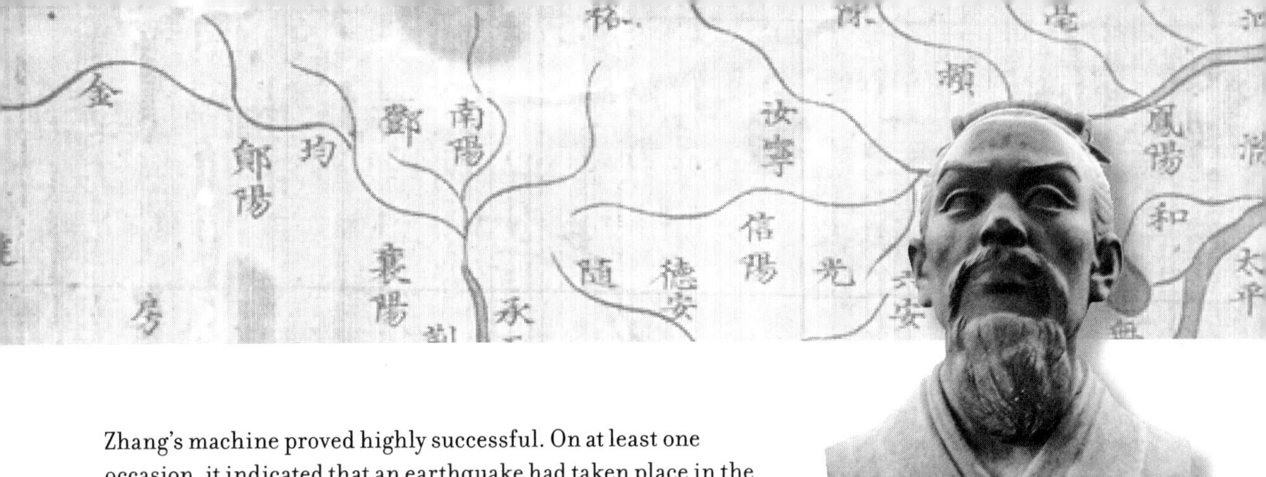

Zhang's machine proved highly successful. On at least one occasion, it indicated that an earthquake had taken place in the east. Since no one felt the tremors, Heng's enemies in the court presumed his machine had failed. But when news of the quake came a few days later, it validated the remarkable genius of a now-fabled scientist.

Zhang believed that earthquakes were caused by the wind: "The chief cause of an earthquake is air, an element naturally swift and shifting from place to place." When the wind was driven into the ground, he believed, it led to the tremors that his machine could detect.

Zhang calculated a more accurate value for pi than anyone in China had done before, and is also credited with inventing an early odometer. He catalogued more than 2,500 stars, more than any contemporary scientists in Greece or Rome. (He, of course, was completely unaware of them, and they of him.) When it came to stargazing, he was part scientist and part astrologist, looking in the sky for favorable portents for the emperor.

820

EASY AS 1, 2, 3

Science by the numbers

It is very likely that you aren't familiar with the name of Muhammad ibn-Musa Al-Khwarizmi. But the work he did more than a thousand years ago affects your life every day—count on it.

Al-Khwarizmi was a mathematician, astronomer, and geographer. Born in what is now Uzbekistan, this Arab scholar eventually found a place at the renowned House of Wisdom in Baghdad—perhaps the greatest center of learning in Islam's golden age.

While there, Al-Khwarizmi wrote a book explaining Hindu concepts in mathematics, including the symbols used in India for counting. Three hundred years after his death, the book was translated from Arabic into Latin. It helped introduce a radical new way to count and do math—using what Europeans called Arabic numerals: 1, 2, 3, 4, 5, 6, 7, 8, 9, and 0. These were much easier to use than Roman numerals, especially for division and multiplication. They eventually came to be used throughout the world.

If that wasn't enough, Al-Khwarizmi wrote another book, called *The Compendius Book on Completion and Balancing.* In Arabic, that's *Kitab al-Jabr wal-Muqabala.* Building on the work of earlier scholars, he laid out a revolutionary way of doing calculations, which has become one of the main branches of mathematics—and the bane of many a high school student.

Take a closer look at that title, particularly the words *al-jabr.* Or, as we know it today: Algebra.

Al-Khwarizmi's book also introduced European mathematicians to the Hindi concept of zero, unknown until that time in the West. Zero comes from the Arab word *sifra*, meaning "empty."

Al-Khwarizmi's name was translated from Arabic into Latin as Algoritimi, the basis for the word "algorithm."

It took a while for Arabic numerals to catch on. Al-Khwarizmi's book was translated by Abelard of Bath around 1100. One hundred years later, the Italian mathematician Fibonacci wrote an influential book advocating use of the new numbers. But even a hundred years after that, in the 1300s, merchants in Florence were forbidden to use Arabic numerals.

1010

FLIGHT BEFORE WRIGHT

Before Orville and Wilbur, there was Eilmer

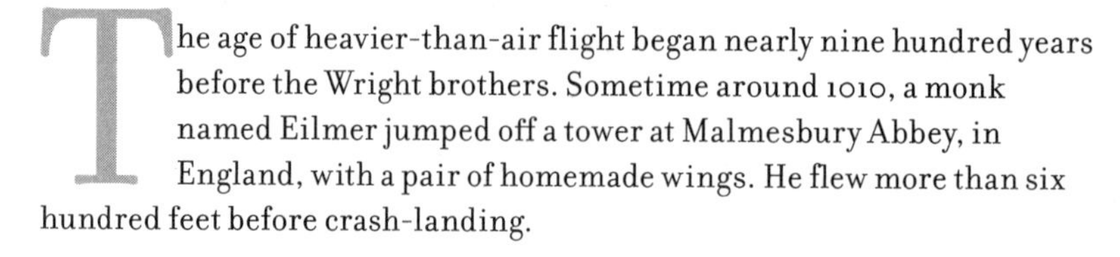

The age of heavier-than-air flight began nearly nine hundred years before the Wright brothers. Sometime around 1010, a monk named Eilmer jumped off a tower at Malmesbury Abbey, in England, with a pair of homemade wings. He flew more than six hundred feet before crash-landing.

That might seem like the stuff of myth, but the feat was recorded just a few years later by a historian named William of Malmesbury in his book *Gesta regum Anglorum* (*History of English Kings*), considered the finest historical work of its day. William was also a monk at the same abbey and knew some of Eilmer's contemporaries. His levelheaded account rings true:

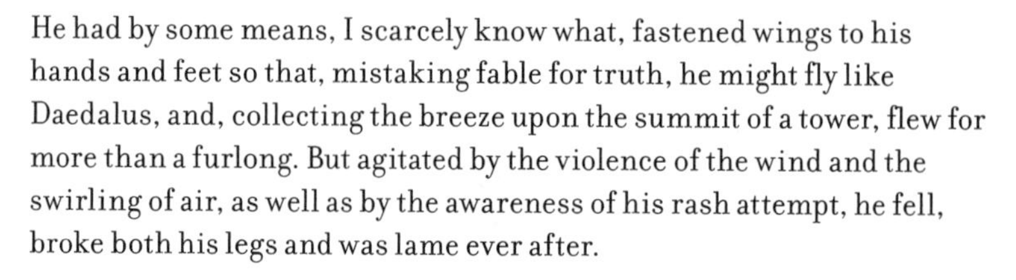

> He had by some means, I scarcely know what, fastened wings to his hands and feet so that, mistaking fable for truth, he might fly like Daedalus, and, collecting the breeze upon the summit of a tower, flew for more than a furlong. But agitated by the violence of the wind and the swirling of air, as well as by the awareness of his rash attempt, he fell, broke both his legs and was lame ever after.

Eilmer was a good enough aeronautical engineer to realize the reason for his rocky landing.

> He himself used to say that the cause of his failure was his forgetting to put a tail on the back part.

Armed with this knowledge, Eilmer planned to fly again with an improved apparatus—but the abbot forbade him. The future of flight would have to wait.

In the year 1260, English philosopher and scientist Roger Bacon, who knew the story of Eilmer's pioneering effort, speculated about the future of manned flight. "Flying machines can be constructed so that a man sits in the midst of the machine revolving some engine by which artificial wings are made to beat the air."

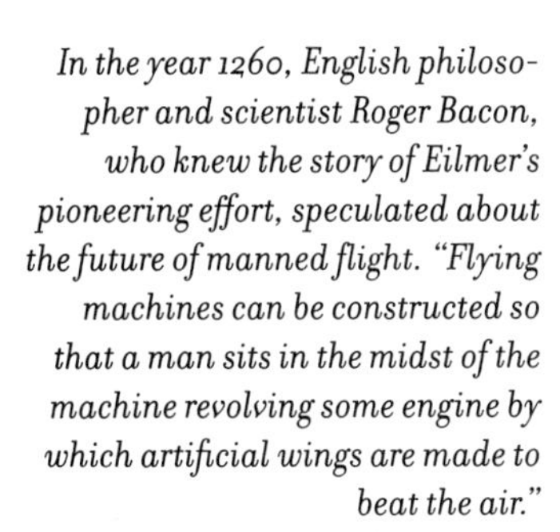

Before they built their powered aircraft, the Wright brothers perfected their aeronautical concepts in gliders. These were made of wood, cloth, and wire, and are simple enough that they could easily have been built centuries before. It is thought-provoking to speculate how the history of flight might have been different if some determined genius had become inspired to follow up on Eilmer's example.

There is now a stained-glass window at the Malmesbury Abbey, portraying the monk Eilmer.

1608

LIPPERSHEY'S LOOKER

How could we forget Hans?

Hans Lippershey's name isn't exactly a household word today, but it was on everyone's lips when the Dutch inventor unveiled his visionary creation in October of 1608. Just two days after he petitioned the parliament of the Netherlands for a patent, a committee was appointed to examine his invention. This examination was considered so important it was to take place in a turret of the royal mansion! Just two days after that, the committee reported that the invention worked so well they had offered Lippershey 900 florins (about $15,000 in today's money) to make one for the government. Fast action, with fame and fortune seemingly certain to follow.

The handy item Lippershey brought into the world was called a *kijker*, or, in English, a "looker." We would call it a telescope.

The spectacle-maker's request for a patent was ultimately denied, in part because members of parliament thought it would be too easy to copy. They were right. Many people started making their own, including a math professor in Italy. Soon the professor was making even more powerful telescopes. Then he turned the world upside down with a simple gesture.

He pointed his telescope to the heavens.

He observed stars, planets, and moon in greater detail than anyone had ever seen them. His observations proved that the earth revolved around the sun, and launched a new era of astronomy. And so Hans Lippershey is relegated to the trivia drawer, while the man who walked in his footsteps is forever remembered: Galileo.

Lippershey was a German-born spectacle-maker living in the Netherlands. After he filed for a patent, several other Dutch inventors came forward to say that they had invented it first. This was also cited as a reason for not granting the patent, although none of the other claimants seem to have brought forward an actual telescope to substantiate their claims.

Galileo's observation of four moons circling Jupiter made it clear to him that everything in the universe didn't revolve around the earth. Further observations convinced him that Copernicus was right, that the earth and planets revolved around the sun. In 1633, the Catholic Church tried him for heresy and forced him to recant, but the genie was out of the bottle, and eventually all the world came to share Galileo's point of view.

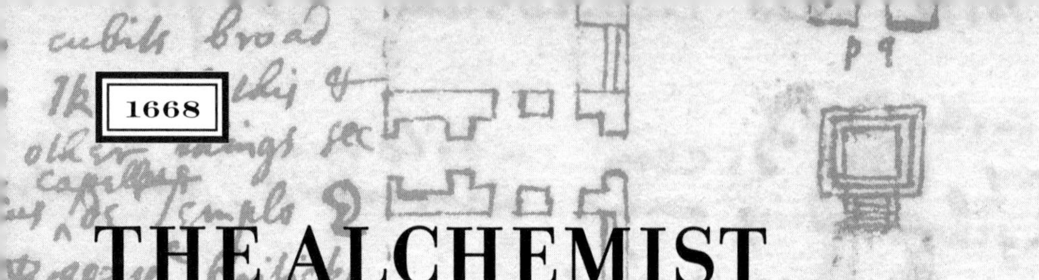

1668

THE ALCHEMIST

Dabbling with the dark side

In his midtwenties, he began his research into alchemy, that mysterious intersection of magic and science, which developed into a lifelong fascination with the occult. He would spend weeks on end searching for a "philosopher's stone" that would transmute base metals into gold. But for him it was more than a material pursuit. "Alchemy tradeth not with metals as ignorant vulgars think," he wrote. He sought the secrets of the universe.

Alchemy was not his main interest, however. He immersed himself in the Bible, analyzing Scripture to tease out its secrets. He taught himself Hebrew so he could read the ancient prophets in their own tongue. He was systematic in his approach, drafting what he called "Rules for interpreting ye words & language in Scripture." He constructed chronologies of the future based on Bible prophecies, predicting that the "tribulation of the Jews" would end in 1944, and that a thousand-year peace would be ushered in around 2370. Here, too, he sought the secrets of the universe.

We do not remember him for either of these lifelong quests, but rather for a third: his explorations into the realm of science. Once again, he sought the secrets of the universe, but *here* he helped uncover them. The work of this alchemist and Bible scholar changed our understanding of the world and made him one of the most celebrated scientists in history:

Isaac Newton.

In the words of Alexander Pope: "Nature and Nature's laws lay hid in night: God said, 'Let Newton be!' and all was light."

Later in life, Newton also added law enforcement to his résumé. In 1696, he was appointed Warden of the Mint. He threw himself into the battle against counterfeiters, personally leading manhunts and conducting interrogations.

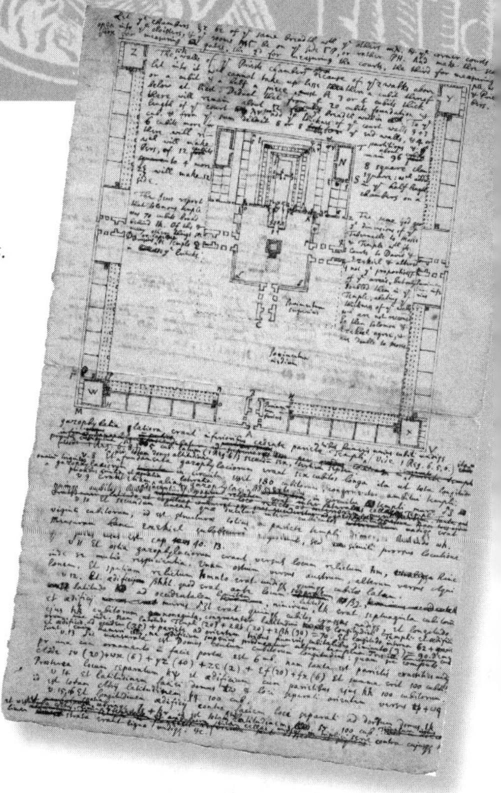

Newton analyzed the Old Testament to discover clues about the dimensions and layout of the Temple of Solomon, then painstakingly created a detailed floor plan. He believed that the dimensions of the temple were themselves sacred, and would help unlock other secrets of the Bible.

That old myth about how Newton saw an apple fall from a tree and "discovered" gravity? There's a seed of truth in it. Newton told several of his associates that seeing an apple fall from the tree in the garden led him to contemplate how far gravity extended. Did it reach all the way out to the moon and the planets? These musings led him to years of work that created the mathematical foundation for understanding gravity.

1715

SYBILLA MASTERS

The first lady of American invention

Sybilla Masters was the wife of a prominent Philadelphia merchant. She was also America's first female inventor—at least the first one we know of. The reason we know of her inventiveness is because she was the only colonial woman to obtain a patent from the King of England.

In 1712, the entrepreneurial Mrs. Masters invented a new way to pulverize corn into meal. Figuring that this was going to be the next big thing, she was determined to protect her idea. Pennsylvania didn't offer patents at the time, but that didn't deter her.

She set sail for England.

It took Sybilla several years to get what she came for. In the meantime, she opened a London hat store to support herself. What Londoners thought of this energetic female inventor from across the pond is largely unrecorded. But on November of 1715, the British government issued a patent to Thomas Masters for "a new invention found out by Sybilla, his wife, for cleaning and curing the Indian corn growing in the several colonies in America."

A telltale mark of the times—even though it was *her* idea, and *her* years of effort that got the patent—the law required that it be issued in her husband's name.

Before she returned to Philadelphia the following year, she was granted another patent: for a new way of making palmetto hats.

Hats off to Sybilla Masters—a colonial housewife with a passion for inventing.

Thomas Masters set up a mill using his wife's invention to make a kind of cornmeal she called Tuscarora rice. An aggressive marketer, Sybilla claimed that it could help cure tuberculosis. Alas, for all her efforts, it failed to become a big seller on either side of the Atlantic.

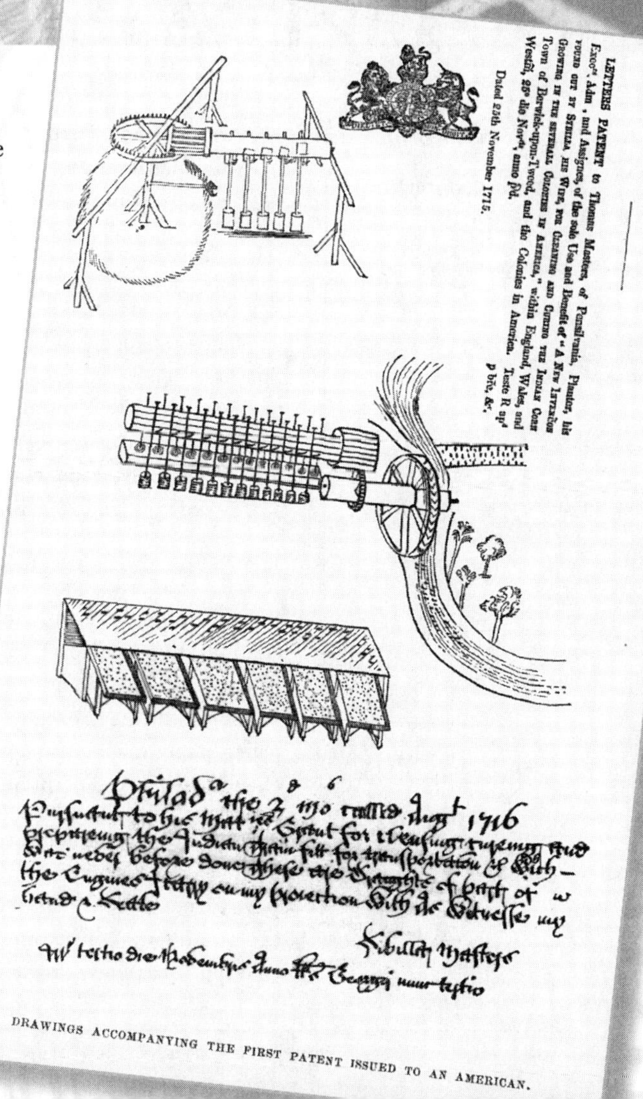

Americans !

courage your own Manufactori

and they will Improve.

1719

PAPER TRAIL

A walk in the woods that made paper plentiful

In the 1600s, Europe was hit with a crippling shortage. People had to deal with the fact that a valuable commodity was in increasingly short supply. What was it?

Rags.

Rags were used to make paper, and paper was in great demand. Publishers of books, newspapers, and political pamphlets—all clamored for more paper. But there just weren't enough rags. Advertisements appeared, asking women to "save their rags." In 1666, England banned the use of cotton and linen for the burial of the dead, decreeing they must be saved for making paper. One entrepreneur even suggested using the cloth from Egyptian mummies. The scarcity of rags led to fearful paper shortages in Europe and America.

Then a French scientist took a walk in the woods.

René-Antoine Ferchault de Réaumur was an accomplished physicist and chemist. He was also a man who loved bugs. Walking in the woods one day, he came upon an abandoned wasp nest. Delighted, he began to examine it in detail, and an astounding fact dawned on him: the nest was made of paper, paper made by wasps, paper made *without* the use of rags. How? By chewing wood and plant fibers.

What wasps could do, he argued, man could find a way to do also. It took decades, but his discovery was the spark that inspired inventors to develop ways to make paper from wood pulp. Thanks to Réaumur's nature walk, we can now do what once would have been considered almost criminal: crumple up a piece of paper and throw it out.

Americans !

Encourage your own Manufactories, and they will Improve.

Ladies. ſave your RAGS.

AS the Subſcribers have it
in contemplation to erect a PA-
PER-MILL in *Dalton*, the enſuing
ſpring: and the buſineſs being very ben-
eficial to the community at large, they
flatter themſelves that they ſhall meet with
due encouragement. And that every woman, who has the good of her country,
and the intereſt of her own family at
heart, will patronize them, by ſaving her
rags, and ſending them to their Manu-
factory, or to the neareſt Storekeeper—
for which the Subſcribers will give a generous price.

HENRY WISWALL,
ZENAS CRANE,
JOHN WILLARD.

Worceſter. Feb. 8, 1801.

The invention of rag paper is credited to a Chinese eunuch named T'sai Lun in the year 5 CE. Until then the Chinese were writing on pieces of bamboo, which was awkward, or scrolls of silk, which were very expensive.

Paper is so plentiful today that the average office worker uses ten thousand sheets a year—and thinks nothing of it. Of course, the flip side of that is deforestation—every year an area of tropical rain forest the size of North Carolina is cut down—in part, to meet the world's paper needs.

1743

BEN THE WEATHERMAN

The answer was blowing in the wind

Ben Franklin was the most famous scientist of his day. The man who captured lightning in a jar was also the inventor of the lightning rod, the Franklin stove, bifocals, and even an improved urinary catheter. His inquiries ranged into every area of science. One of them helped open a door into the future of weather forecasting.

On October 21, 1743, Franklin hoped to see a lunar eclipse that was supposed to take place at 8:30 p.m. But by late afternoon a northeast wind began to blow, and a violent storm blanketed Philadelphia for the next twenty-four hours.

The storm was a big one, and newspapers from different cities reported the damage up and down the coast. When Franklin read one of the Boston papers, however, it wasn't the storm damage that caught his eye—it was an account of the eclipse. He checked with his brother, who lived in Boston, and confirmed that the storm didn't reach Boston until at least an hour after the eclipse.

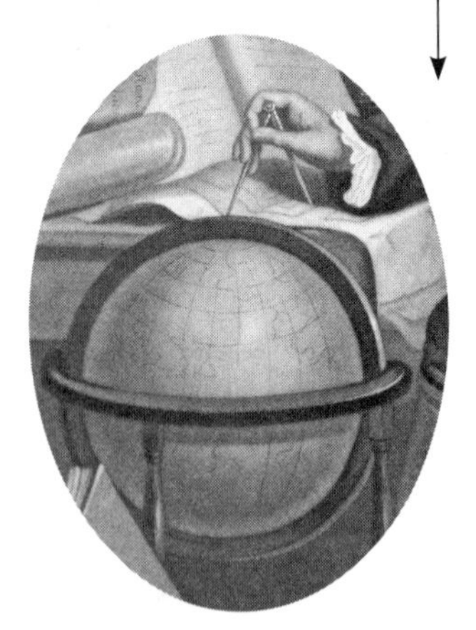

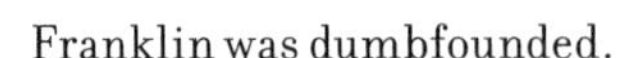

Franklin was dumbfounded.

He and most other people believed that the big storms known as nor'easters traveled from northeast to southwest, since that's the direction the wind blew. Now he knew that the storms traveled in the exact opposite direction. Armed with this information, he went on to develop theories about storm tracks and pressure systems that proved remarkably accurate.

More than 150 years later, geographer William Morris Davis wrote that Franklin's keen-eyed observations "began the science of weather prediction."

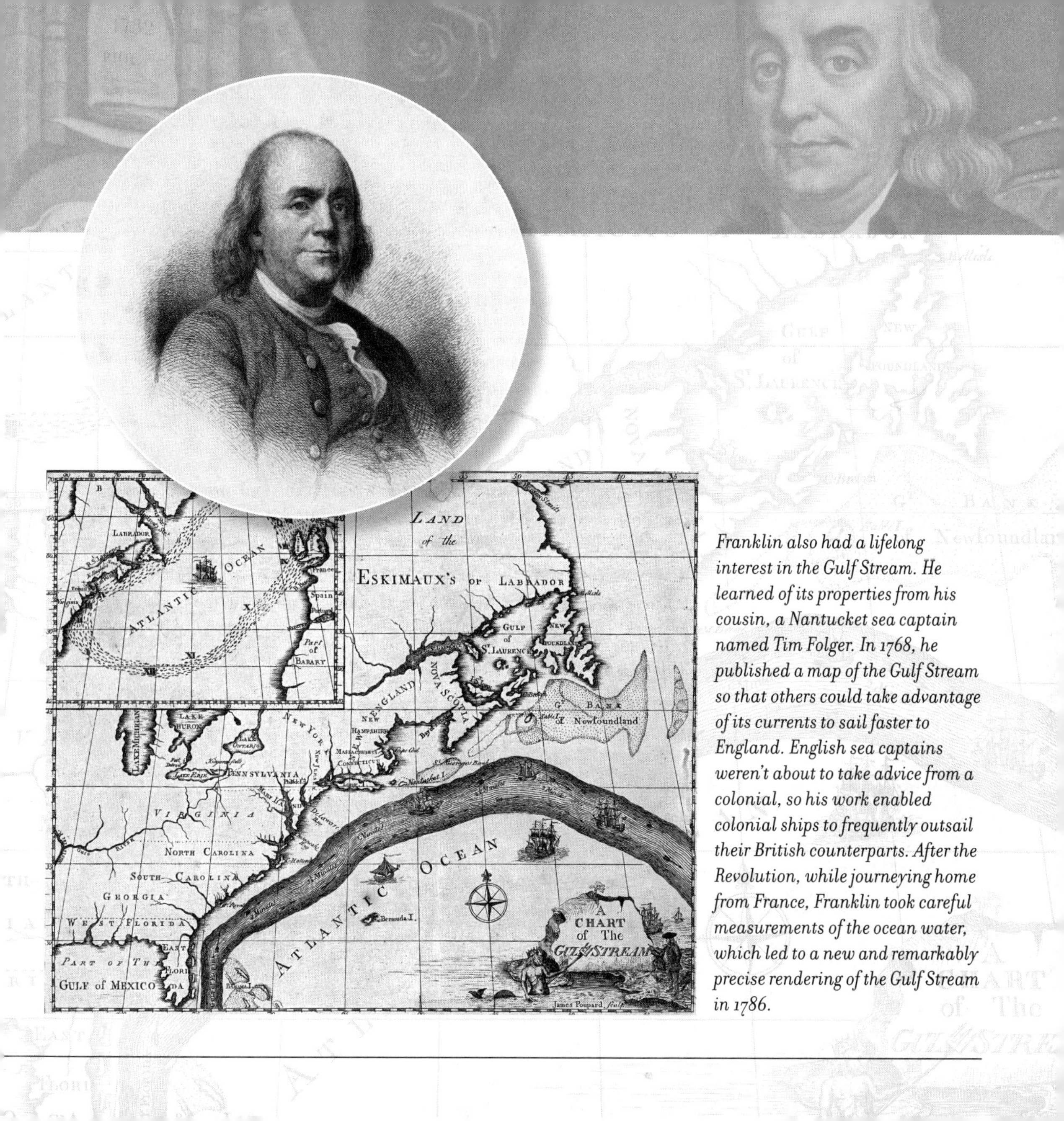

Franklin also had a lifelong interest in the Gulf Stream. He learned of its properties from his cousin, a Nantucket sea captain named Tim Folger. In 1768, he published a map of the Gulf Stream so that others could take advantage of its currents to sail faster to England. English sea captains weren't about to take advice from a colonial, so his work enabled colonial ships to frequently outsail their British counterparts. After the Revolution, while journeying home from France, Franklin took careful measurements of the ocean water, which led to a new and remarkably precise rendering of the Gulf Stream in 1786.

1746

RING OF FIRE

A Paris experiment that was a real shocker

The scene on this particular April day was surreal: two hundred Carthusian monks, all in their robes, each holding one end of a twenty-five-foot wire connecting him to the next monk. They formed a giant human chain nearly a mile long, which snaked across the grounds of the Grand Convent of the Carthusians on the outskirts of Paris.

The chain looped back around so that the monk at the end stood next to the monk at the beginning. Then the monastery's abbé, a celebrated French scientist named Jean-Antoine Nollet, used a special apparatus to charge up a Leyden jar with a load of static electricity. As crowds of people looked on with anticipation, Nollet had the monk at one end grasp the metal plate on the outside of the jar. The monk on the other end touched the metal pole at the top.

That completed the circuit, discharging the static electricity through the monks.

All two hundred monks reacted at the same moment to the electric shock. Some jumped in the air, others shouted, all dropped the wires between them. The demonstration was a crowd pleaser, but it also offered dramatic proof of two points that were to have a huge impact on the world: electricity could travel long distances, and it could do so instantaneously.

Electricity would one day tie the entire globe together in a web of instant communications. And, in a very real way, our wired world was born on that April day outside Paris.

Nollet performed a similar experiment for King Louis XV, sending electricity through 180 of his palace guards.

One of France's most famous scientists, Nollet engaged in a bitter transatlantic argument with Ben Franklin about the nature of electricity. In the end, Franklin's ideas won wide acceptance and Nollet's were discredited.

1769

FIRST CAR

Before Ford, Mercedes, and Ferrari, there was Cugnot

The first automobile was built far earlier than you might think. In fact, it took to the streets of Paris almost a hundred years before Henry Ford was even born.

French engineer Nicholas-Joseph Cugnot invented the first self-propelled road vehicle in 1769. Powered by a steam engine, his *fardier à vapeur* ("steam wagon") could achieve a top speed of only 2.5 miles per hour. And it could only operate for about twenty minutes before it literally ran out of steam. It was nonetheless quite powerful; Cugnot had served in the French army, and his vehicle was designed to haul artillery.

The following year, Cugnot produced a second vehicle, another three-wheeler that could carry four passengers. In 1771, he drove one of his machines into a stone wall, making him not only the inventor of the first automobile but also the first person to get into a car accident.

Cugnot's machines might have paved the way for a transportation revolution coming at about the same time as the American Revolution. But his sponsor at the French court fell out of favor, and so did Cugnot. Cars would have to wait.

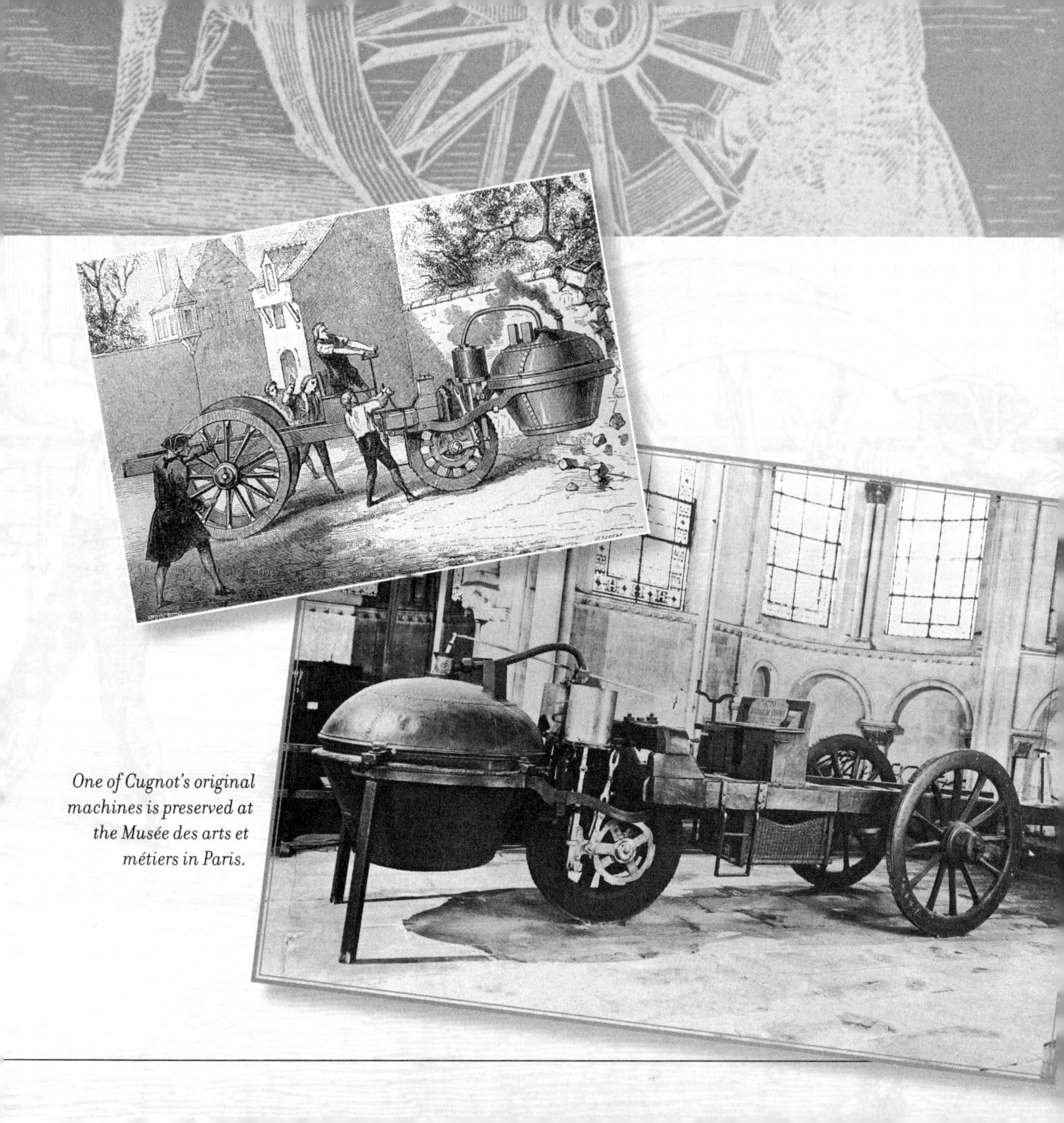

One of Cugnot's original machines is preserved at the Musée des arts et métiers in Paris.

1781

THE MUSIC MAN

The happy consequences of picking up a new instrument

In March of 1781, a backyard stargazer in Bath, England, spotted a comet with one of his homemade telescopes. William Herschel wasn't a professional astronomer; he was a talented and well-known musician. He composed symphonies, conducted orchestras, and played the cello, the oboe, and the organ.

Now, with a new instrument, he was about to redraw the heavens.

Nervously, since he was only a hobbyist, he reported his sighting to the scientific community. His report caused quite a stir. Not so much because of the comet itself—others were able to verify the sighting. But because Herschel casually mentioned that he was using a telescope with a magnification power of 2000X. To other scientists, that sounded crazy. After all, the finest telescope in Britain's Royal Observatory only had a magnification of 270X. Herschel, they thought, must be some sort of crank.

No, a virtuoso.

Herschel, without anyone realizing it, had become the finest telescope-maker in the world. And that comet he spotted? It turned out to be a new planet, the first discovered since antiquity: Uranus. The discovery rocketed him to international fame. He was appointed personal astronomer to King George III, and went on to build the biggest telescope in the world. He devoted the next forty years to a systematic study of the skies, revolutionizing astronomy and making the most of his extraordinary second career.

Herschel called the new planet the Georgian Star, in honor of King George. French astronomers called it Herschel, in honor of its discoverer. It was a German astronomer who suggested calling it Uranus. In Roman mythology, Uranus is the father of Saturn and the grandfather of Jupiter. For years, the planet was known by all three names, until Uranus eventually won out.

The German-born Herschel recruited his sister Caroline as his assistant, and she became a renowned astronomer in her own right, discovering multiple comets and winning several notable astronomy awards.

Herschel constructed more than four hundred telescopes. The most famous was a reflecting telescope, financed by King George III, which was the biggest telescope in the world at the time it was built. The king and the archbishop of Canterbury came to visit during construction, when the telescope's forty-foot-long barrel was still on the ground. The king invited the bishop to walk through it: "Come, my Lord Bishop, I will show you the way to Heaven!"

1793

THE MECHANICAL INTERNET

How networking came to Europe more than two hundred years ago

In the 1790s, dozens of odd-looking towers sprouted up across France. They comprised the backbone of a new, state-of-the-art communications network.

The towers were the brainchild of a French inventor named Claude Chappe. Each one had mechanical arms that could be rotated into ninety different positions, visible ten miles away. An operator would set the arms in a certain position. Then the operator in the next tower would see that configuration through a telescope, and set his tower's arms in the same position. Messages rippled down the line at more than one hundred miles an hour, astonishing for the time.

Chappe wanted to call his new invention the tachygraph, meaning "fast writer." But a friend convinced him to give it a different name, meaning "far writer." And so he called it . . . the telegraph.

When Napoleon seized power in 1799, he was quick to grasp the military advantage of high-speed communications. He ordered new lines built out from Paris in every direction. Eventually, more than five hundred towers connected France's major cities. Other countries followed suit, and lines of towers snaked across Europe, Russia, and even North Africa.

The invention of the electrical telegraph eventually made the Chappe telegraph obsolete. Today only a few of the hundreds of towers remain . . . silent monuments to the world's first high-speed network.

Chappe began his experiments during the French Revolution, which caused him no end of trouble. Angry mobs twice destroyed his equipment, because they suspected he was trying to communicate with jailed royalty.

Like many inventors, Chappe was hounded by others who claimed they deserved the credit for his ideas. Eventually, he became so depressed that he took his own life by throwing himself down a well.

1799

THE OTHER BEN

Tinker, Tailor, Soldier, Spy—and lots more

He grew up poor outside of Boston, but went on to become a Bavarian count. A spy for the British during the American Revolution, he was later offered the chance to start the U.S. Military Academy. His scientific research made him famous, but then again, so did his soup. Once, he was as celebrated as Ben Franklin; today, few have heard of him.

Meet Benjamin Thompson.

He married a rich widow in Concord, New Hampshire (then known as Rumford), but his loyalist activities forced him to flee the colonies in 1776. After serving in the British government, he ended up going to work for the Duke of Bavaria. He reorganized the Bavarian army, reforming everything from the kitchens to the way soldiers dressed. He also established workhouses for the poor. For a while, he even ran Bavaria. The prince made him a count, and he took the name Count Rumford.

His wide-ranging scientific inquiries started when he was a teenager. His work on gunpowder and artillery helped launch the science of ballistics. His research into heat laid the groundwork for modern thermodynamics. He invented the coffee percolator, the kitchen range, a new and improved fireplace, and thermal underwear.

To feed the poor, he created a simple soup that he hoped would alleviate world hunger. Oddly enough, it became one of his most celebrated achievements. "Who is ignorant," said a speaker at his

POWDER

funeral, "of what service he has rendered to humanity by introducing the general use of the soups which go by his own name."

Today almost everyone is ignorant of Count Rumford and his many achievements.

The "Rumfordized fireplace" reshaped the hearth and chimney to make them far more efficient. Some of its features remain a part of modern fireplaces.

The count left money in his will for Harvard to establish a professorship in his name. In 1856, the third Rumford professor, Eben Horsford, invented the new and improved baking powder that remains in many kitchen cabinets today.

In the early days of the Revolution, Rumford spied on his fellow colonists for British General Thomas Gage. President John Adams was unaware of this in 1799, when he offered Rumford the chance to come back to America to create a new military academy. Although he had solicited the offer, Thompson turned it down, possibly because the British government threatened to embarrass Thompson and the United States by revealing his earlier espionage work. As it was, his spying did not become public until a century after his death.

1803

SHELL SHOCK

The revolutionary weapon that changed warfare forever

At the start of the 1800s, a new weapon appeared on the battlefield.

It was the brainchild of an English officer who spent thirty years perfecting it. A hollow artillery shell was filled with smaller musket balls, along with a charge of gunpowder ignited by a fuse. The shell could be launched long-distance at the enemy's lines. When it exploded in midair, it spread a deadly carpet of metal shards over a wide area.

The inventor of the shell devoted all his free time to perfecting it, and poured his life savings into the project. The British army finally adopted the shell in 1803, and first used it in the Napoleonic wars. It proved frighteningly lethal on massed troops. It so terrified French soldiers that they believed the British had poisoned their cannonballs.

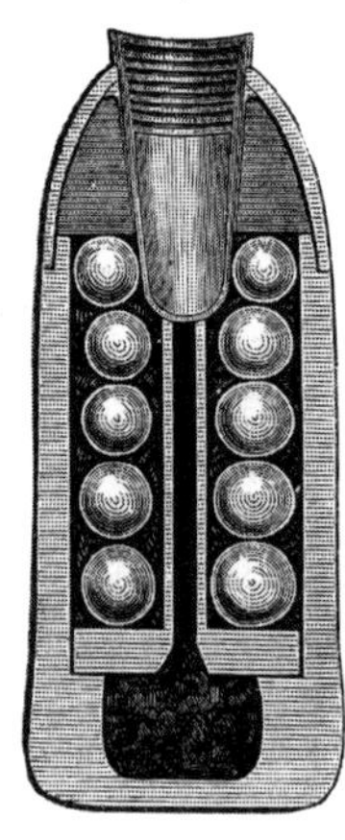

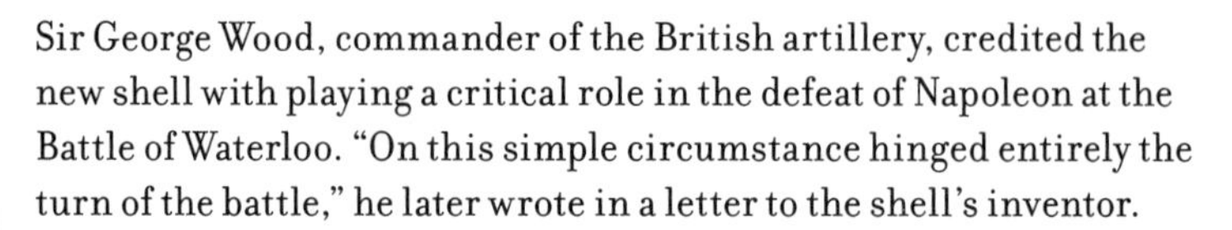

Sir George Wood, commander of the British artillery, credited the new shell with playing a critical role in the defeat of Napoleon at the Battle of Waterloo. "On this simple circumstance hinged entirely the turn of the battle," he later wrote in a letter to the shell's inventor.

Artillery became infinitely more terrifying, and the name of the officer who invented the shell became a household word . . .

Henry Shrapnel.

The Duke of Wellington ordered the new shell kept secret, even though he knew it meant denying Shrapnel fame and honor for his invention. The shells were not referred to by Shrapnel's name until decades later.

Shrapnel's shells were the "bombs bursting in air" that Francis Scott Key saw during the bombardment of Fort McHenry in the war of 1812.

Napoleon ordered unexploded British shells to be disassembled so he could fathom their secrets—but he never managed to duplicate them.

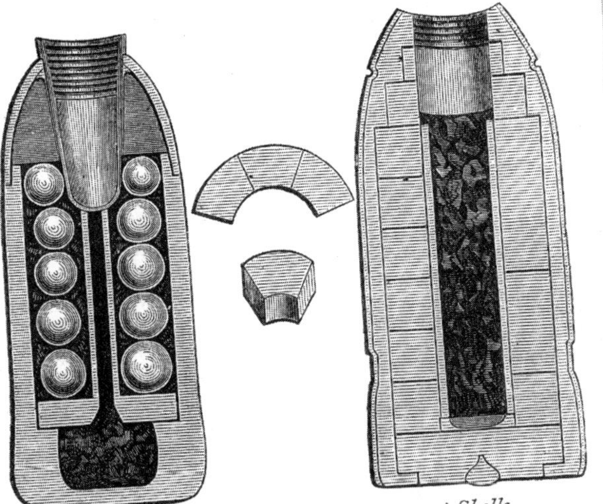

The Shrapnel and Segment Shells.

1805

FIRST CAR, USA EDITION

Automotive ingenuity, American-style

What was the first automobile invented in America? That honor may go to a vehicle with the wonderfully unlikely name of the Orukter Amphibolos, built in Philadelphia in 1805. (The name means "amphibious digger.")

Its inventor, Oliver Evans, had a contract to build a steam dredge to help clear the Schuylkill River. But Evans, a talented inventor who had been interested in building a horseless carriage for decades, used the money to build a vehicle that has been described as America's first steamboat, first locomotive, and first car. Nearly thirty feet long and weighing seventeen tons, it was powered by a revolutionary high-pressure steam engine that was Evans's own invention. The engine could drive a paddlewheel when the boat was afloat, or shift power to the wheels when it was on land.

In July of 1805, it apparently lumbered around in Philadelphia's Centre Square, where Evans charged onlookers admission until it eventually broke down. And Evans claimed in the newspaper to have navigated it down the Schuylkill. Accounts vary, and the Orukter may have worked better in Evans's imagination than it did on the roads.

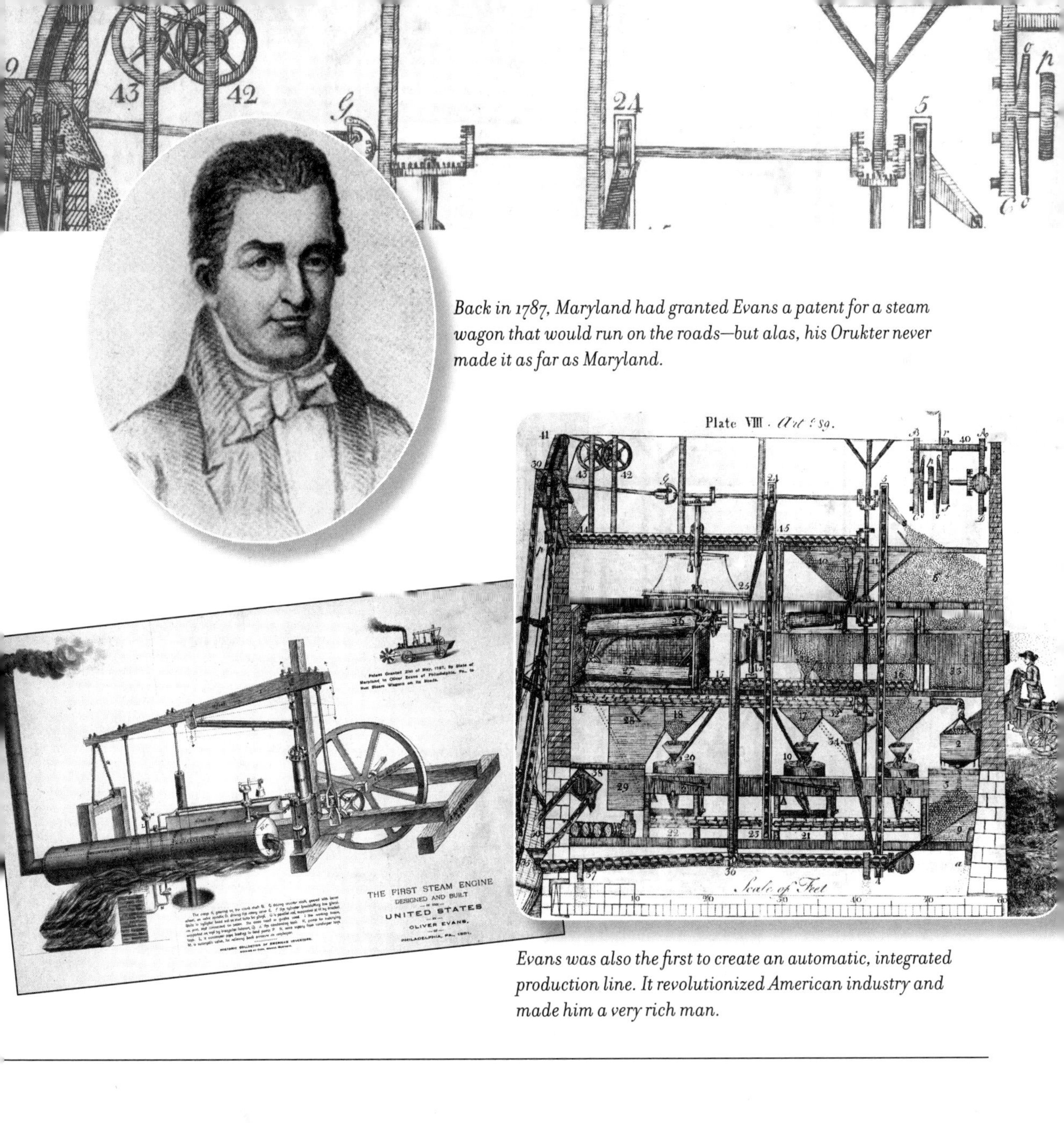

Back in 1787, Maryland had granted Evans a patent for a steam wagon that would run on the roads—but alas, his Orukter never made it as far as Maryland.

Evans was also the first to create an automatic, integrated production line. It revolutionized American industry and made him a very rich man.

1809

CAN DO!

A revolution in France creates a revolution in food preservation

It is an odd fact that the can was invented about fifty years before the can-opener. (That's a long time to wait for dinner!) On the other hand, the process we call canning predates the use of the can.

Confused yet?

In the wake of the French Revolution, France found itself fighting just about every other country in Europe. The government of France offered a prize to any inventor who could create a better way to preserve food in order to supply the military in the field.

It took him more than a decade of experimentation, but Parisian distiller and chef Nicholas Appert won the twelve-thousand-franc prize in 1809, with a process that became known as "appertizing." Food was sealed in champagne bottles (how did they get it through that small hole?) and then heated in boiling water. The heating killed the germs, but Appert didn't know that—he just knew it worked. The food was easy to transport and wouldn't spoil for months or years.

Across the English Channel, a man named Peter Durand borrowed (i.e., stole) the idea and obtained a patent—with a notable improvement. He wanted to use metal containers. The first cannery in England opened soon afterward, putting food in *can*isters made out of *tin*plated steel.

Tin cans.

They were generally opened with knives or hammers, until the first can-opener was invented in 1858. And then, to quote a phrase often employed by my father:

"The tin can bended, and the story ended."

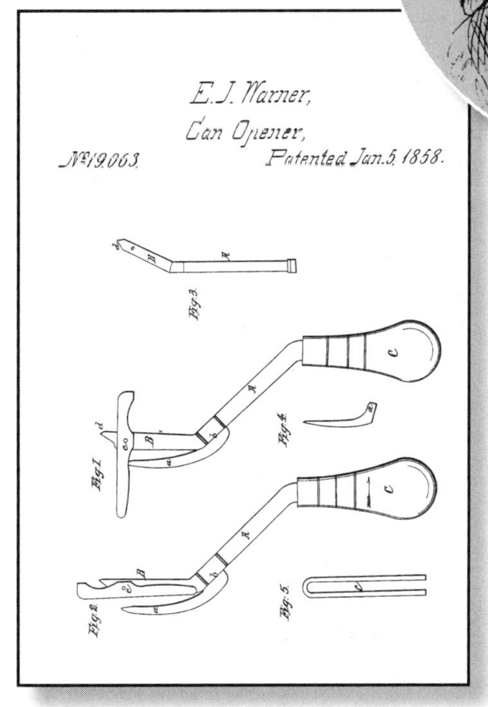

Ezra Warner of Waterbury, Connecticut, is credited with patenting the world's first can-opener in 1858. He called it an "Instrument for Opening Cans." The allure of building a better can-opener has remained strong. Literally hundreds of ideas have been patented over the years for a device to accomplish this seemingly simple task.

Appert opened the world's first cannery (although he used bottles, at first) in 1804. It remained in operation for more than a century. He is also the inventor of the bouillon cube.

1816

BAH-BUMP GOES THE STETHOSCOPE

Who would ever think modesty could be the mother of invention?

In September of 1816, a buxom young woman paid a visit to a French physician named Dr. René Laennec. He felt certain that she had heart problems. But the morals of the day prevented him (a bachelor) from listening to her heart in the normal fashion, by putting his ear to her chest. Besides, trying to hear a heart through such an ample bosom would be a trifle difficult.

The doctor was also a musician. He rolled a tube from paper—like a flute—and he touched this to his patient. "I was surprised and gratified," he wrote, "at being able to hear the beating of the heart with greater clearness than ever before." Soon after, he fashioned a hollow cylinder out of wood with a funnel at one end—the first stethoscope.

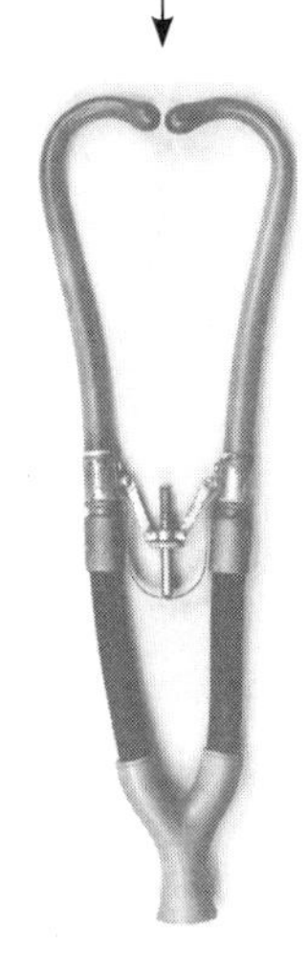

Many doctors were quick to adapt to this new technology. But some resisted. One American doctor put it this way: "He that hath ears to hear, let him use his ears and not a stethoscope."

But the new device allowed doctors to diagnose problems of the heart and lungs they had never been able to uncover before. A new window on illness was opened.

Thanks to modesty.

In 1852, a New York doctor named George Camman invented the modern stethoscope, with a bell-shaped chest-piece and a tube running to each ear. Medicine's most ubiquitous instrument has changed very little since then.

René Laennec thought his invention was such a simple device that it didn't need a name. But when others pushed him for a name, he suggested "stethoscope"—from the Greek word stethos, *meaning "chest,"* and skopos, *meaning "observer."*

1822

FIRST COMPUTER

The first computer: invented more than fifty years before the lightbulb

Railroads were still brand-new in 1822 when a British mathematician named Charles Babbage dreamed up the idea of a mechanical calculator. He called it a "difference engine." Like the railroad, it was designed to operate by steam.

It took Babbage ten years to build just one section of his difference engine. When it was completed, people marveled at its sophistication. But even as he was working on the difference engine, he conceived of something even more ambitious: a machine that could be programmed with punch cards to perform even the most complex calculations.

In other words, a mechanical computer.

Babbage made hundreds of drawings outlining his new idea. He envisioned that this "analytical engine" would be about as big and heavy as a small locomotive, containing thousands of finely machined gears. It would be able to add or subtract forty-digit numbers in a few seconds, and multiply similarly large numbers in about two minutes.

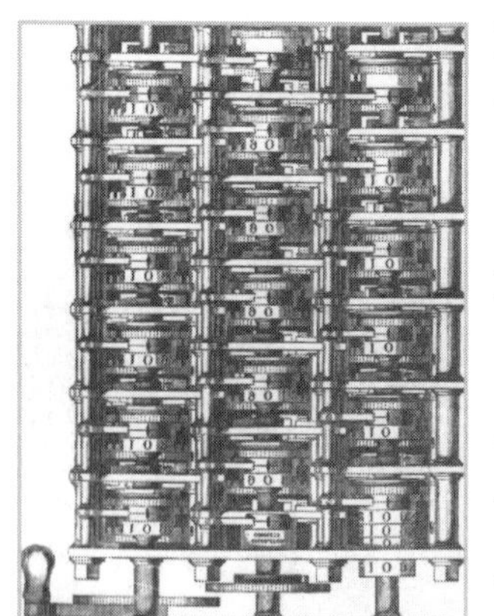

Babbage was a bit ahead of his time, since this was more than a hundred years before the first electronic computer was created. Technology was not advanced enough to create the complex machine. He did, however, develop other, more practical things, including something for those newfangled railroad locomotives.

You can still see it on the front of old steam engines.

The cowcatcher.

Babbage was such a perfectionist that every time he started to build the analytical engine, he came up with an idea that would render that version obsolete. Thus it became more and more complex, until it became just about impossible to build.

Babbage was assisted by Ada Lovelace, the daughter of the famous romantic poet Lord Byron. She was a talented mathematician, and is celebrated as the first computer programmer, since she wrote programs for the analytical engine. Babbage called her the "Enchantress of Numbers."

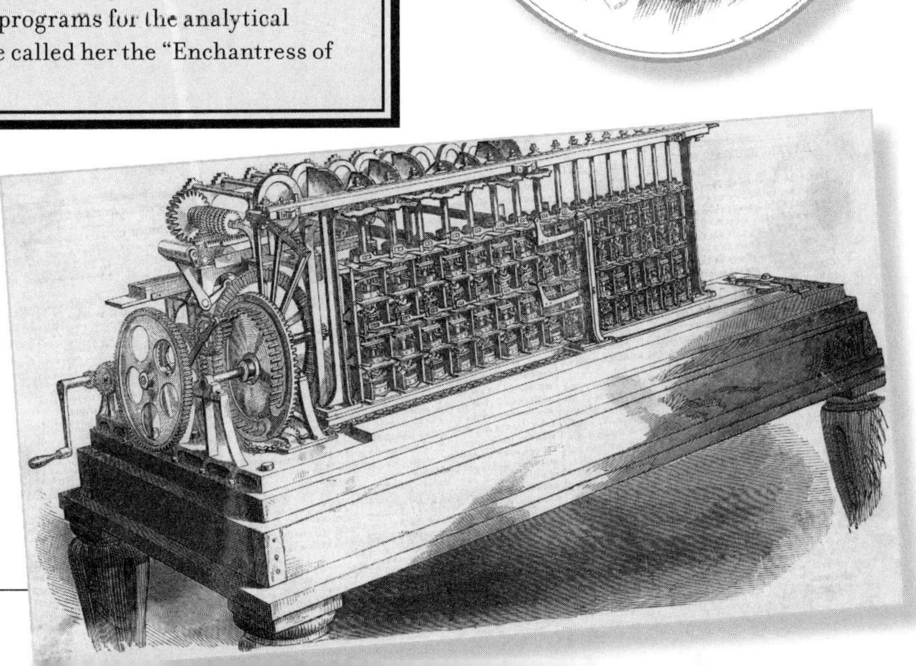

1826

PICTURE THIS

When his son joined the army, it led to the world's first photo

Joseph Nicéphore Niépce was a retired military man caught up in the latest fad sweeping France: lithography. It involved creating images in oil on a piece of limestone, then chemically treating them so that the limestone could be used to print copies. Joseph wasn't much of an artist himself, so he let his son Isidore draw the pictures while he handled the delicate chemical process required to make it work.

But when seventeen-year-old Isidore joined Napoleon's army in 1814, his father needed some other way to create images on the limestone. Joseph hit upon the idea of shining light through a pre-existing artwork onto a light-sensitive coating on the limestone.

It sounds complicated, and, in fact, he was never able to do it. But it led him to a decade of experiments trying to capture images on light-sensitive substances. Eventually, that work culminated in the world's first photograph, taken from an upstairs bedroom of his house. He called it a *heliograph*, Greek for "sun drawing."

So why have most of us never heard of Joseph Nicéphore Niépce?

By the time he made his heliograph, he was in his early sixties and in ill health. A few years later, he reluctantly partnered with a younger man, an ambitious Paris painter named Louis Daguerre.

Daguerre perfected the process and introduced the daguerreotype in 1837, four years after Niépce's death. So it was Daguerre who reaped the fame for enabling people to capture themselves and their world on film.

Niépce's first photograph involved the use of a pewter plate coated with bitumen of Judea, a kind of asphalt that hardens when exposed to light. It required an eight-hour exposure.

Photography wasn't Niépce's only invention. In 1807, he and his brother Claude patented the Pyréolophore, one of the world's first internal-combustion engines. It had a piston and cylinder system similar to twentieth-century car engines. Claude spent years—and most of his fortune—in a futile attempt to interest the world in their invention.

Making photography practical required reducing exposure time. Daguerre's key breakthrough was a result of serendipity. He exposed a copper plate covered with silver iodide. When no image appeared, he put it away in a cupboard. Opening the cupboard a few days later, he found a picture on the plate. He eventually discovered that mercury vapor from a broken thermometer had brought out the image. By accident, Daguerre had invented the darkroom.

1831

A NATURAL SELECTION

Choosing the right traveling companion can change history

At age twenty-five, Robert FitzRoy became captain of a British navy ship after the previous captain committed suicide. He understood the crippling loneliness that a long voyage could inflict on a ship's commanding officer. So when he set out on another long voyage, to survey the waters off South America, he sought out a companion to accompany him. Someone he could talk to, someone who shared his love of science, someone who would use the trip to collect animal and plant specimens from the places they would visit.

The post was offered to an accomplished naturalist named Leonard Jenyns, and he initially agreed. But after agonizing about it, he changed his mind. Jenyns was also a clergyman—that was how he supported himself—and he didn't feel he should abandon his parishioners.

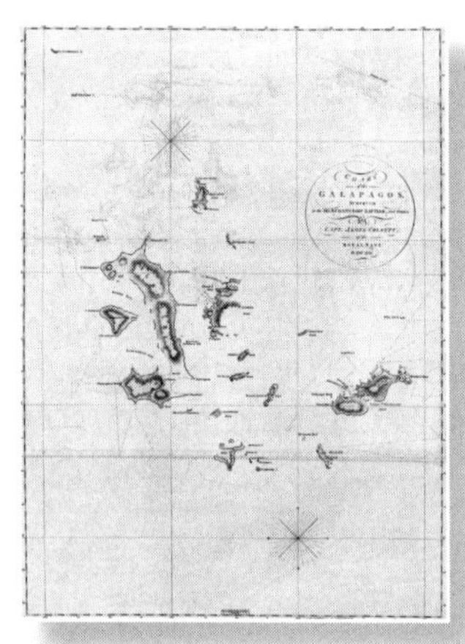

So FitzRoy sought out another candidate. The position was offered to another young naturalist, who came highly recommended by Jenyns. But he, too, refused, and for a remarkably similar reason: his father wanted him to go into the clergy, and this trip was a waste of time that would interfere with his son's future career.

But the second-choice candidate put a full-court press on his father, who eventually assented to the voyage. And that's how Charles Darwin embarked on the H.M.S. *Beagle* in 1831 for a five-year voyage that would lead him to write his book *On the Origin of Species* and develop the theory of evolution.

Robert FitzRoy rose to the rank of admiral, served as governor of New Zealand, and became a pioneering meteorologist. He was the first head of the British Meteorological Office, where he coined the term "weather forecast," and initiated the practice of sending out daily forecasts for publication in the newspaper.

Darwin's father, a doctor and poet, didn't think his son was likely to amount to much. "You care for nothing but shooting, dogs, and rat-catching," he once shouted at his son, "and you will be a disgrace to yourself and all your family."

After meeting Darwin, FitzRoy almost rejected him—because of the shape of his nose! "He doubted whether anyone with my nose could possess sufficient energy and determination for the voyage," wrote Darwin years later. "But I think he was afterwards well satisfied that my nose had spoken falsely."

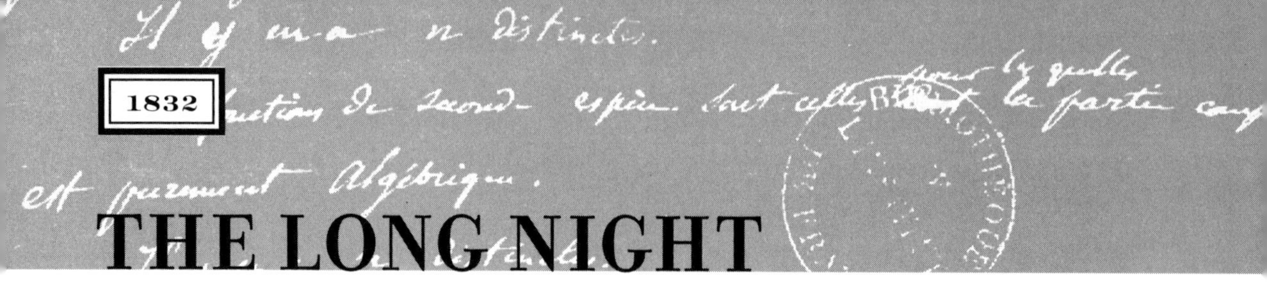

THE LONG NIGHT

Last will and testament of a mathematician

When a man knows he is to be hanged in a fortnight," said English author Samuel Johnson, "it concentrates his mind wonderfully." For French mathematician Evariste Galois, it was a duel in the morning that concentrated his mind on the night of May 29, 1832.

Galois had earned a reputation as a brash genius—a man with talent and a temper. Then he became entangled in a duel. (Its cause remains a mystery.) Anticipating his possible death, he spent the night pouring out his thoughts in a letter to his friend Auguste Chevalier. He worked against the clock to summarize his life's work in mathematics and lay out new theorems that he hadn't had time to prove. He labored feverishly to complete his thoughts before morning. "I have not time, I have not the time," he scribbled in the margin of the letter.

The next day, Galois went out to fight the duel. Shot in the intestines, he suffered terribly for two days before dying. His work, however, lived on. After his death, he was hailed for developing a branch of algebra known as group theory. It turned out to play an important role in quantum mechanics, the means by which scientists seek to understand the universe at a subatomic level.

One can't help but wonder what else Galois might have accomplished if he had lived longer. The biggest tragedy of his life was its brevity.

Evariste Galois was just twenty years old when he died.

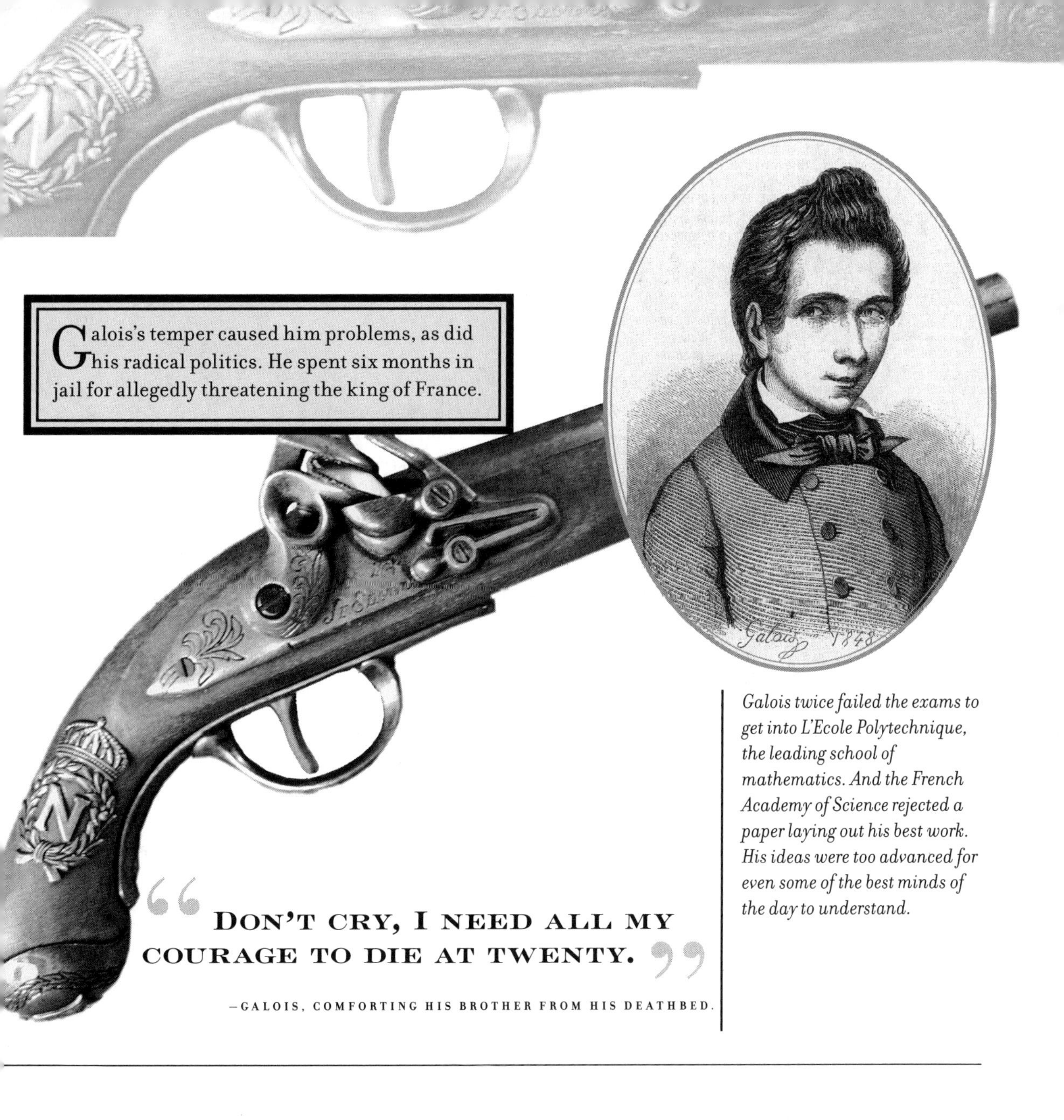

Galois's temper caused him problems, as did his radical politics. He spent six months in jail for allegedly threatening the king of France.

Galois twice failed the exams to get into L'Ecole Polytechnique, the leading school of mathematics. And the French Academy of Science rejected a paper laying out his best work. His ideas were too advanced for even some of the best minds of the day to understand.

"DON'T CRY, I NEED ALL MY COURAGE TO DIE AT TWENTY."

—GALOIS, COMFORTING HIS BROTHER FROM HIS DEATHBED.

1842

OF RAILROADS AND TRUMPETS

An experiment that set a serious tone

Oh, to be an innocent bystander alongside the tracks of the Rhine railroad in the Netherlands that day in 1842. Along came a train with an open car full of trumpeters blowing with all their might. Another group of musicians could be seen on the platform of the train station, ears cocked, furiously scribbling down notes as the train pulled in. What in heaven's name was going on?

It was an ingenious experiment conducted by Dutch scientist Diederik Buys Ballot to test a brand-new theory proposed by Austrian scientist Christian Doppler: the Doppler effect.

Doppler was the first to theorize that light waves and sound waves appear to change in frequency if the source is moving toward or away from the observer. He suggested a formula for calculating what the change should be. Buys Ballot set out to test the idea.

The trumpeters were told to hold a G note. As the train pulled into the station, fourteen musicians there estimated the change in frequency, measuring it in eighths or sixteenths of a note. The experiment confirmed Doppler's theory.

The Doppler theory is now a cornerstone of many different fields, including radio astronomy, radar, and medical imaging. The trumpeters on the train car are forgotten, but their song plays on.

Also known for his work in meteorology, Buys Ballot founded the Royal Dutch Meteorological Institute and remained its director until his death in 1890.

1844

THE ELASTIC MAN

The obsession that changed the world

At the age of thirty-two, Charles Goodyear fell in love with rubber. His hardware business had gone bankrupt, and he was deeply in debt. He read about the new waterproof gum from Brazil, and found himself fascinated by its potential.

But Goodyear was discovering rubber just as the rest of the world was becoming disenchanted with it. Rubber in the 1830s had a big flaw. It became rigid in cold weather, and melted when it was hot. Rubber products had seen a brief flurry of popularity, but now most people had given up on the stuff.

Goodyear set out to change that. He commenced years of experimentation, mixing rubber with everything imaginable to make it more viable. When he ran out of money, he pawned his wife's jewels. His family was reduced to poverty, and he shuttled in and out of debtor's prison. When his son died of malnutrition, he couldn't even afford to buy a coffin for him. Still he persevered.

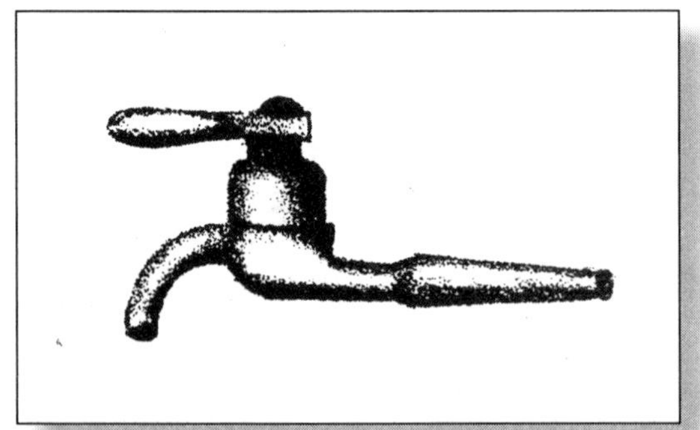

By 1839, he had been experimenting for five years, with little to show for it. One day, he was working with some rubber that was mixed with sulfur and lead. He accidentally dropped it on the stove. When he removed it, he discovered that the charred piece of rubber had hardened, but still remained supple.

Goodyear had found the answer: a process to strengthen rubber that became known as vulcanization. It would soon make rubber an integral part of all our lives, thanks to a man obsessed.

Goodyear never got rich from his invention. He eventually sold the patent rights. When he died in a boardinghouse in 1860, he was $200,000 in debt.

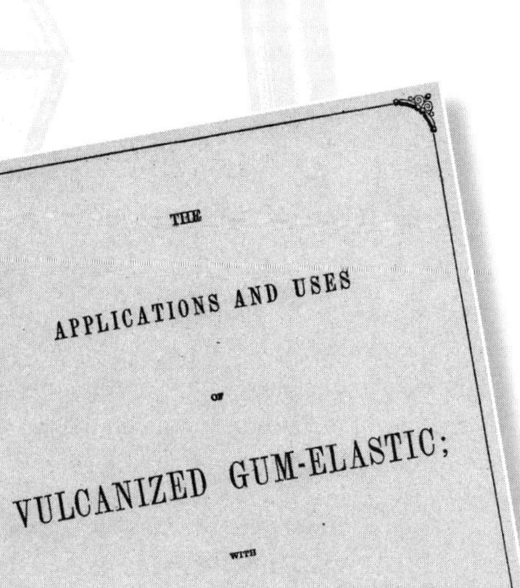

THE

APPLICATIONS AND USES

OF

VULCANIZED GUM-ELASTIC;

WITH

DESCRIPTIONS AND DIRECTIONS FOR MANUFACTURING PURPOSES

BY CHARLES GOODYEAR.

VOL. II.

NEW HAVEN:
PUBLISHED FOR THE AUTHOR.
1853.

Goodyear wrote a book with page after page of ideas for things he thought could be made out of rubber: suspenders, dollar bills, telescopes, and pianos among them. Oddly enough, one thing that didn't make the list was rubber tires.

1849

LINCOLN'S IMP'd MANNER OF B

PATENT PRESIDENT

Yankee ingenuity from a country lawyer

President Abraham Lincoln is famous for wearing his stovepipe hat. He used to keep his papers in it. But he also wore another hat most people don't know about—that of an inventor. Lincoln is the only president ever to hold a patent.

In his youth, Lincoln worked as a boatman on the Mississippi River. In the late 1840s, Lincoln journeyed to and from Washington via Great Lakes steamship while he was serving in Congress. On one occasion, the steamer got stuck on a sandbar. All the passengers and cargo had to be unloaded to float the ship over the sandbar. It was a long and tedious process.

This gave Lincoln an idea: What if there was a way to float the boat over the sandbar without having to empty it out and reload it? When he got back to his law office in Illinois, he began working on a device that would do just that. He whittled the model in between court appearances while talking to his law partner about how it would revolutionize shipping on the nation's inland waters. The result: Patent #6469, "Manner of Buoying Vessels."

Lincoln's invention proved impractical, and was never manufactured. Still, it stands out as a unique achievement. As *Scientific American* opined in 1860: "It is probable that among our readers there are thousands of mechanics who would devise a better apparatus for buoying steamboats over bars, but how many of them would be able to compete successfully in the race for the Presidency?"

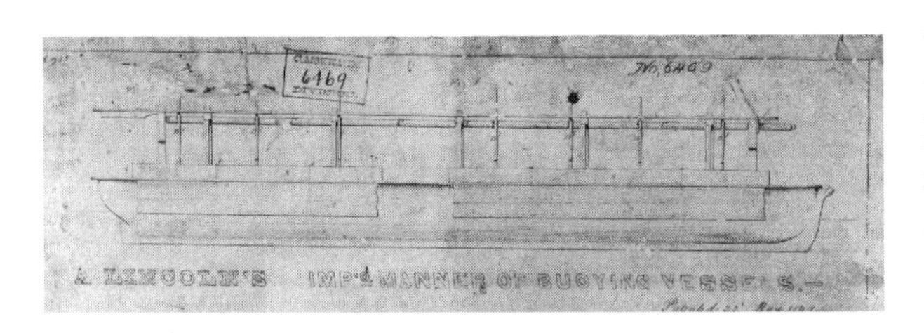

"ALTHOUGH I REGARDED THE THING AS IMPRACTICABLE I SAID NOTHING, PROBABLY OUT OF RESPECT FOR LINCOLN'S WELL-KNOWN REPUTATION AS A BOATMAN."

—WILLIAM HERNDON, LINCOLN'S LAW PARTNER

Lincoln's idea was to add "adjustable buoyant air chambers" below the water line. When the ship was stuck, these could be filled with air to float it off. It is remarkably similar to the idea that inventors later used to help create the modern submarine.

1854

GOING UP

The shocking demonstration that led to the birth of the skyscraper

The scene was the Crystal Palace Exposition in New York. Inventor Elisha Otis was speaking to a crowd from a platform suspended high above by a rope. Suddenly, the rope was severed with the swift cut of an ax.

Screams and gasps could be heard as the platform began to fall. But after a drop of only a few inches, Otis yelled, "All safe, gentlemen, all safe!"

Elisha Otis did not invent the first elevator. He invented something more important . . . the elevator brake. That meant his elevators were different from all that came before: they were safe. Otis had sold only three of his elevators the year before the demonstration, but afterward, sales began to shoot up.

And so did elevators.

Because of his invention, modern buildings would soon begin to leap toward the sky. The development of a safe elevator helped make possible the skyscraper, and the profile of the American cityscape would be changed forever.

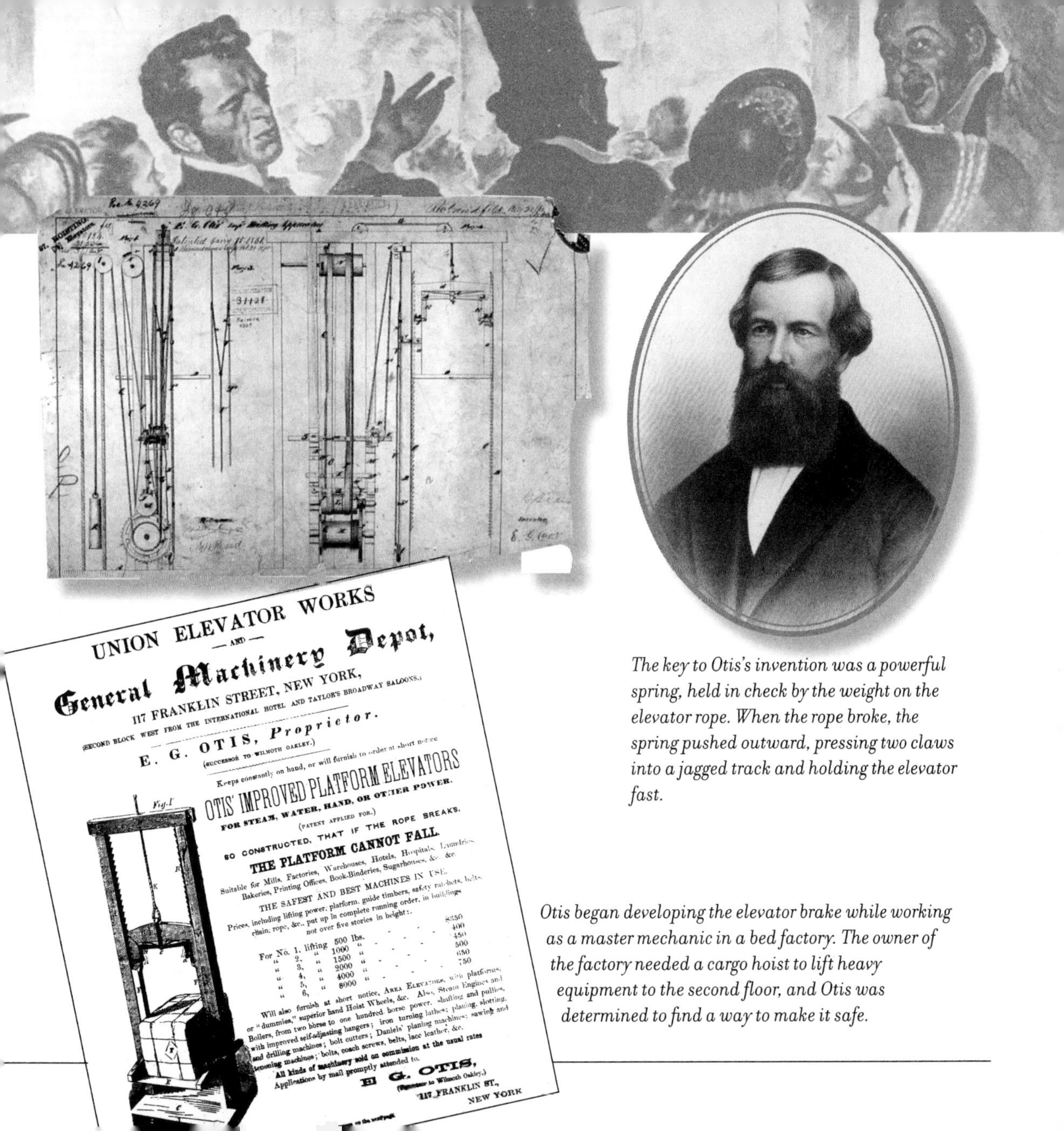

The key to Otis's invention was a powerful spring, held in check by the weight on the elevator rope. When the rope broke, the spring pushed outward, pressing two claws into a jagged track and holding the elevator fast.

Otis began developing the elevator brake while working as a master mechanic in a bed factory. The owner of the factory needed a cargo hoist to lift heavy equipment to the second floor, and Otis was determined to find a way to make it safe.

BIGO'S BEETS

The French distiller who inspired a war on disease

Bigo was a man with a problem. He had a business distilling alcohol from beets in Lille, France. But something was wrong with the fermenting vats. Their yield smelled like sour milk and had to be thrown out. It was killing his business, and he didn't know what to do.

Bigo's son was studying at a local university, and suggested that his father consult with one of his teachers. Bigo was doubtful that an academic could help, but desperation overcame skepticism. He got in touch with the thirty-four-year-old chemistry professor and asked for his help. Little did he know that his request would mark a turning point not only for his business but for the health of all mankind.

The professor's name: Louis Pasteur.

Pasteur gathered samples from good and bad batches. He spent hour upon hour examining them. "He is now up to his neck in beet juice," wrote his wife to her father. Pasteur put some of the spoiled fermentation juice under a microscope and saw large numbers of rod-shaped things moving around that weren't in the healthy vats.

Scientists had seen germs before, but thought they were an unimportant byproduct of disease. Pasteur realized that these tiny rods were alive, and *causing* the problems. He advised Bigo to use a microscope to select only healthy yeast, and the problem cleared up.

Pasteur's discovery that germs are living organisms that cause disease launched the field of microbiology and led to remarkable breakthroughs in the prevention and treatment of disease. Can't beat that.

One of those who read of Pasteur's work with interest was a Scottish surgeon named Joseph Lister. If micro-organisms caused infections, he reasoned, doctors should treat wounds with antiseptic and make sure their hands and instruments were sterile. His findings, published in 1867, revolutionized surgery and dramatically lowered the number of deaths caused by infection. The mouthwash Listerine, developed in 1879, is named after him.

"In the fields of observation, chance favors only the prepared mind."

—LOUIS PASTEUR

Pasteur's theories were greeted at first with disbelief, and took decades to be fully accepted. His astonishing ability to get results eventually squelched all doubters. He developed the process of pasteurization, found a cure for rabies, and saved France's silk industry when silkworms were threatened by disease.

1858

DEM DRY BONES

The beginnings of a new dinosaur age

In the summer of 1858, hardly anyone had ever heard of dinosaurs. The word was a new one, coined to describe some large bones found in England a few years before. And certainly no one yet had an accurate view of what a dinosaur really looked like. But that was all about to change, in the unlikeliest of places:

New Jersey.

William Parker Foulke was a Philadelphia lawyer spending his summer a few miles out of town in the suburb of Haddonfield, New Jersey. One of his neighbors regaled him with a tale of some mysterious large bones that workmen had found twenty years earlier in a clay pit. Foulke had a keen interest in science, and was greatly intrigued. He pinpointed the spot, now overgrown with trees and brush, and organized a dig. Ten feet under the surface they started finding what Foulke called "monstrous bones."

Lots of them.

Foulke hadn't just uncovered a few bones. He had found the largely complete skeleton of a dinosaur, the first one ever discovered. In honor of its finder, it was named *Hadrosaurus Foulki*, literally, "Foulke's Big Lizard." It rocked the scientific world, offering irrefutable proof that dinosaurs as big as houses once ruled the earth.

A few years later, the reconstructed dinosaur skeleton was put on display at the Academy of Natural Sciences in Philadelphia. It created a sensation—nobody had seen anything like it. The exhibit sparked a national fascination with the huge creatures of earlier epochs, which has never diminished since.

The reconstructed skeleton stood twenty-six feet long and thirteen feet high. Scientists examining the skeleton determined that it probably walked upright, which was the first indication that some dinosaurs walked on two legs and not four. One scientist referred to it standing upright like a kangaroo, and it became widely known as the "Great Kangaroo Lizard."

Foulke was an ardent prison-reformer and abolitionist.

The discovery of dinosaur bones in New Jersey brought together two young paleontologists, Edward Drinker Cope and Othniel Marsh, who later became fierce competitors in the so-called Bone Wars of the late 1800s. Each mounted many fossil-collecting expeditions in the West, but also resorted to bitter attacks and underhanded methods to undermine the other. Ultimately, the effort ruined both—but left the world with a wealth of dinosaur bones.

1860

TELEPHONE TALE

Being first doesn't always mean being famous

Alexander Graham Bell invented the telephone in 1876—right? Well, actually a German schoolteacher managed to do it fifteen years before Bell, but he didn't have very good PR.

In 1860, Philip Reis rigged up what he described as an "artificial ear." This crude instrument was built from an improbable assortment of items: a violin, a knitting needle, an ear carved in wood, even a piece of sausage. "I succeeded in inventing an apparatus by which . . . one can reproduce sounds of all kinds at any desired distance . . . I named the instrument [the] 'telephone.' "

It worked—but poorly. So Reis kept tinkering with it. He hooked up a wire between his workshop and the school, which convinced his students that he was using the telephone to eavesdrop on them. He worked on his telephone for several years, and his models became more and more sophisticated.

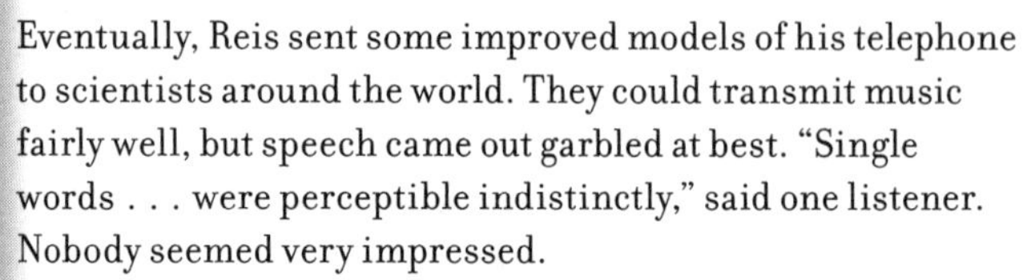

Eventually, Reis sent some improved models of his telephone to scientists around the world. They could transmit music fairly well, but speech came out garbled at best. "Single words . . . were perceptible indistinctly," said one listener. Nobody seemed very impressed.

Most scientists at the time regarded the Reis telephone as little more than a toy. Crushed, he abandoned his work and died of tuberculosis in 1874. Two years later, Alexander Graham Bell filed the patent that earned him the glory and financial rewards of telephone invention.

Alexander Graham Bell didn't set out to invent a telephone. He was trying to build a better telegraph that could send multiple messages down the same wire. But once he realized he could make a telephone, he could think of nothing else. Bell's great competitor in telephone invention was Elisha Grey, whose application for a telephone patent was filed just hours after Bell's. At the time, Grey didn't seem to understand what he'd lost out on. "The talking telegraph is a beautiful thing from a scientific point of view," he wrote in 1876. "But if you look at it in a business light, it is of no importance."

Bell, Reis (at left), and Grey are not the only competitors for the title of the "first to invent the telephone." Other contenders put forward over the years include Italian-American set designer Anthony Meucci and Charles Bourseul, an employee of the French Telegraph service.

1863

BIRTH OF A BEVERAGE

The little-known pedigree of the world's most popular soft drink

In 1863, a Corsican chemist named Angelo Mariani came up with a new drink consisting of wine and coca—the plant that cocaine comes from.

It had a real kick to it.

Vin Mariani became a worldwide sensation. Mariani sold the cocaine-laden drink as a medicinal tonic, advertising it with testimonials from doctors and celebrities that trumpeted the drink's ability to relieve fatigue, dissipate the blues, even invigorate people's sex lives.

Hundreds of imitators soon cropped up. One entrepreneur who set out to copy Mariani's formula for success was an Atlanta pharmacist named John Pemberton. He created a similar cocaine concoction, flavoring his drink with kola nuts and calling it French Wine Cola. Among other claims, he marketed it as a cure for morphine addicts—of which he was one. "A Living Joy to all who use it," his ads proclaimed.

But Pemberton's success with French Wine Cola was quickly tempered by the temperance movement. In 1885, Atlanta voted to go dry—making Pemberton's alcohol-based tonic illegal.

So he re-jiggered the ingredients, taking out the alcohol and adding just the right amount of sweetener. He started selling it to drugstores—as a temperance drink! The change in formula was a recipe for success. And with it came a change in name. From coca leaves and kola nuts:

Coca-Cola.

Pemberton's first advertisements for Coca-Cola described it as "The New and Popular Soda Fountain Drink containing properties of the wonderful coca plant and famous cola nuts."

The cocaine came out of Coca-Cola more than a hundred years ago. The secret formula still uses coca leaves, but they are treated in such a way as to eliminate the drug.

Vin Mariani was a hugely popular product endorsed by doctors, opera singers, President McKinley, and even Pope Leo XIII, who awarded it a gold medal. But the drink faded away in the early twentieth century, as the dangers of cocaine abuse became apparent, and new laws went into effect banning cocaine in many countries,

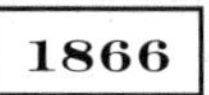

RADIO PROPHET

His ideas were decades ahead of his technology

Who invented radio? The credit is usually shared among a host of famous names: Guglielmo Marconi, Nicola Tesla, Thomas Edison, Heinrich Hertz. But an obscure Virginia dentist named Mahlon Loomis may have beaten them all to the punch.

Loomis was an inventor who had already come up with a new way of making dentures. As early as the 1850s, he began to believe that it was possible to send telegraph signals through the atmosphere without wires.

In October of 1866, Loomis conducted a demonstration that may have been the first-ever wireless transmission, using two kites flown from two West Virginia mountaintops fourteen miles apart. Each kite was covered with copper mesh, and connected to the earth via wire. By connecting and disconnecting the grounding wire on one of the kites, he was able to move a sensitive electrical meter known as a galvanometer, which was connected to the wire on the other kite. He called it "wireless telegraphy" and filed for a patent.

Loomis dreamed of sending telegraph signals across the ocean, but he didn't really understand the science behind his device, and was unable to improve its performance. He died in 1886, broke and discouraged, but still a believer. "I know I am . . . regarded as a crank . . . but I know that I am right and if the present generation lives long enough their opinions will be changed—and their wonder will be that they did not perceive it before."

His equipment may have been imperfect, but his vision was beyond question.

Loomis frequently lectured on his ideas, as he tried to raise money for the "Loomis Aerial Telegraph Company." His increasingly dramatic claims tended to make some wonder if his original transmission had ever really taken place. Evidence suggests that it did.

> **"We may tap the storehouse of the mighty thunder and make it whisper glad tidings over the seas."**
>
> —Dr. Mahlon Loomis

Another forgotten radio pioneer is Kentucky melon farmer Nathan Stubblefield, who was able to send wireless voice transmissions more than a third of a mile in 1902. His system was a little different from radio as we know it—he apparently sent the signals through the ground using induction. But his work, and his vision of centralized stations sending signals out to large audiences, helped to spur public interest in wireless broadcasting. A recluse in his later years, he starved to death in 1928.

1873

CHESTER'S CHAMPIONS

The inventive teenager who didn't muff his big chance

Chester Greenwood's ears got cold when he went ice-skating. So cold it drove him crazy. First they would turn red, then blue. The teenager from Farmington, Maine, tried hats, and scarves, but nothing effectively protected Chester's sensitive protuberances from the arctic assault of the Maine winter.

So he decided to come up with his own solution.

He shaped two pieces of thin wire into loops. Then he asked his grandmother to sew fur covers for them. To keep these ear-insulators from falling off, he designed a springy metal band to connect them. At last his ears were warm and toasty.

At age fifteen, Chester Greenwood had invented earmuffs.

By the time he was nineteen, he began manufacturing his earmuffs. They sold chiefly in Maine in the early years, but soon word of Chester's "Champion Ear Protectors" spread far and wide. A skilled machinist, Chester built the equipment needed to turn them out in large numbers. Within ten years, he was making thirty thousand pairs a year. By the time he died, his company was making ten times that many.

"Build a better mousetrap," said Ralph Waldo Emerson, "and the public will beat a path to your door." Turns out the same thing is true for earmuffs.

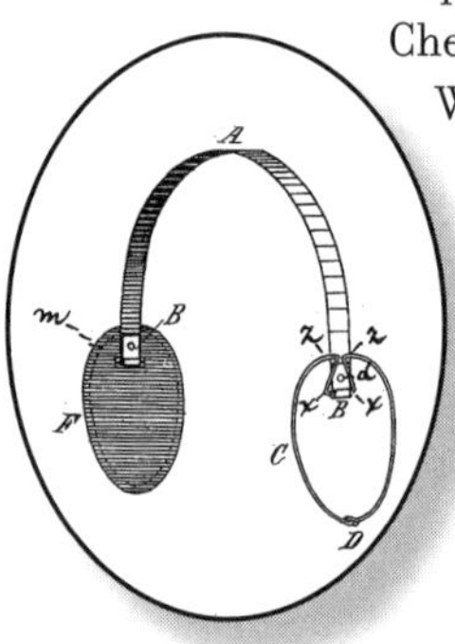

Greenwood lived in Farmington his whole life. He also patented the steel-spring rake, and an improved teakettle.

GREENWOOD'S

"CHAMPION"

EAR PROTECTOR!

An Article has been found at last which will KEEP THE EAR WARM in the Coldest Weather!

Patented in the United States March 13, 1877. Patented in Canada March 12, 1878, and April 6, 1882.

FIG. 1.

FIG. 2.

FIG. 1 represents the CHAMPION EAR PROTECTOR unfolded ready for use.

FIG. 2 represents the CHAMPION EAR PROTECTOR in use.

They Fold so as to be Conveniently Carried in the Pocket!

They are Worn by the Ladies as well as by the Gentlemen!

Every Man, Woman, and Child, should have a pair of these Ear Protectors.

CHESTER GREENWOOD, PATENTEE, WEST FARMINGTON, MAINE, U. S. A.

CHESTER GREENWOOD & CO., Manuf'rs,
WEST FARMINGTON, MAINE.

R. S. McCormick, Bridgetown, Nova Scotia, Manufacturer for the Dominion of Canada.

One advertisement touted the product with this boast: "No disfiguring plastic below the chin to make a man look like an overgrown schoolboy."

Farmington, Maine, celebrates the earmuff inventor on the first Saturday of every December, with "Chester Greenwood Day."

1874

THE DEVIL'S ROPE

The invention that fenced in the West

What invention had the greatest impact on the American West? The six-shooter? The locomotive? The lasso?

Try this one: barbed wire.

In the mid-1800s pioneers headed west in pursuit of the American dream. Attracted by cheap and fertile land, they encountered a thorny problem: how to protect their crops from livestock roaming freely across the range. There wasn't enough wood to make fences. Farmers tried using wire, but the cattle slipped through. Inventors tried adding sharp points to the wire, but a solution that was both practical and effective eluded them.

In 1874, a farmer from DeKalb, Illinois, named James Glidden used a coffee grinder to bend short pieces of wire into sharp, pointy barbs. He strung them out on a smooth wire, then twisted that wire with another one to hold the barbs in position.

Cows hated it. The Indians called it the Devil's Rope. Within twenty years it had effectively closed in the wide-open spaces of the West, ending the days of free-roaming cowboys and cattle drives. Since then it has been threaded across battlefields, enclosed concentration camps, and divided Cold War Berlin.

An amazing legacy for one Illinois farmer.

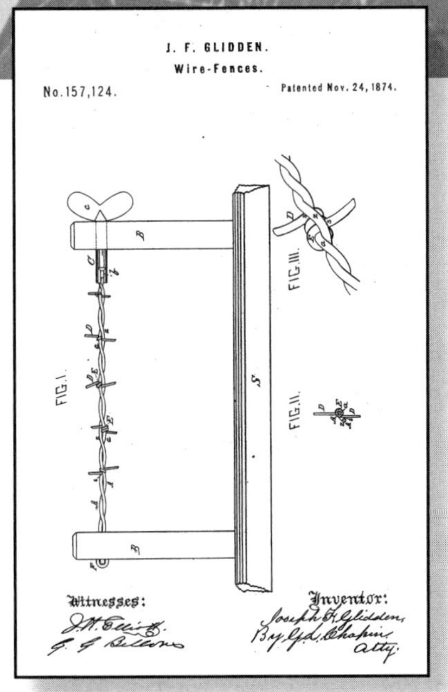

Two other inventors, Isaac Ellwood and Jacob Haish, filed similar patents, but Glidden won a twenty-year patent battle. He was successful both because he came up with a way to lock the barbs onto the wire and because he also invented the machinery to make it. Hundreds of other barbed-wire-related patents have been filed, and today it is estimated that there are over two thousand kinds of barbed wire.

Glidden was inspired to invent barbed wire when he went to a county fair and saw a demonstration of a wire fence holding a board with nails sticking out of it.

The barbed-wire fences of the West inspired the Cole Porter song "Don't Fence Me In."

1879

SWEET AND SOUR

The accidental discovery of artificial sweetener

Constantine Fahlberg was at dinner when he noticed it. Something sweet had gotten on his fingers. Something he was eating, perhaps? No, he realized, this was a taste far sweeter than anything on the table. It must have been something he got on his hands at the lab that day. The next day he began sampling some chemical concoctions he had been working on, and quickly found the sweet stuff: a chemical derived from coal tar. It turned out to be five hundred times sweeter than sugar.

Fahlberg gave it a new name: saccharin. It became the world's first artificial sweetener.

Fahlberg was doing research for a distinguished chemist named Ira Remsen at the time of the discovery, and they announced it together. But then Fahlberg made a secret trip to Washington to claim the patent rights for himself. Remsen, who later became president of Johns Hopkins University, never forgave him. "Fahlberg is a scoundrel," Remsen later said. "It nauseates me to hear my name mentioned in the same breath with him."

Even in its earliest days, some thought saccharin unsafe, and in 1908 the secretary of agriculture tried to have it banned because it was "extremely injurious to health." He might have succeeded except for some blowback from saccharin's biggest fan: President Theodore Roosevelt, who was using it on the advice of his doctors. "Anyone who said saccharin was injurious to his health is an idiot," Roosevelt proclaimed.

And that was that.

Lawyer-turned-cafeteria-operator-turned-teabag-manufacturer Ben Eisenstadt invented the sugar packet in 1947. When he pitched it to Domino's sugar, the company loved the idea, but decided they didn't need Ben or his factory. They began making packets on their own. Ten years later, Ben's son Marv invented a granulated version of saccharin that the two of them decided to market in pink packets. The result was Sweet'N Low (the name comes from a poem by Alfred Lord Tennyson), which came out in 1957, just in time to catch a national diet craze and launch a new generation of low-calorie sweetening products.

SCHLATTER LAB JOURNAL

Accidental discoveries abound in the history of artificial sweeteners. In 1965, chemists at G. D. Searle were working on an anti-ulcer drug. While heating a chemical compound, Dr. James Schlatter spilled some on his fingers. Later in the day, he licked his fingers to pick up a piece of paper, and noticed a sweet taste. At first he thought it was the doughnut he had eaten at his coffee break, but quickly realized his error. Knowing the substance he was working with was safe, he put some in his coffee, and noted the sweet taste in his lab book. The result was aspartame, now the world's No. 1 artificial sweetener.

1879

THE LIGHTBULB BEFORE EDISON

An illuminating history of a bright idea

Thomas Edison is the most celebrated inventor in American history. Among his many inventions, the most legendary is certainly the incandescent lightbulb.

But, in fact, Thomas Edison didn't invent the lightbulb.

More than twenty inventors came up with the idea before him. An English chemist named Joseph Swan lit hundreds of houses in England with his lightbulbs before Edison invented his. Swan always claimed that Edison stole his ideas. Edison solved that problem by going into business with Swan—their joint venture was known in England as Ediswan.

But Swan wasn't the first, either. There were a dozen lightbulb inventors before him, going all the way back to the day in 1802 when British chemist Humphrey Davy demonstrated that passing electricity through a strip of platinum would generate light.

So why does Edison get the credit? For one thing, he and the team at his lab in Menlo Park, New Jersey, invented a better lightbulb than anyone before them. More important, Edison was thinking big. His vision extended far beyond one bulb. He invented a whole illumination system with generators, switches, and fuses that could bring light to a house, a neighborhood, and eventually the entire globe.

ajesty's Colonies and Plantations abroad the "I
OBTAINING LIGHT BY ELECTRICITY" communicated
reigner residing abroad In
A

SIR HIRAM MAXIM

In 1844, three years before Thomas Edison was born, a nineteen-year-old Cincinnati wunderkind named John Wellington Starr invented a lightbulb that featured a carbon filament light inside a vacuum—similar to what Swan and Edison came up with decades later. Starr might have gone on to fame and fortune, but he died of pneumonia before he turned twenty-two. His lightbulb died with him.

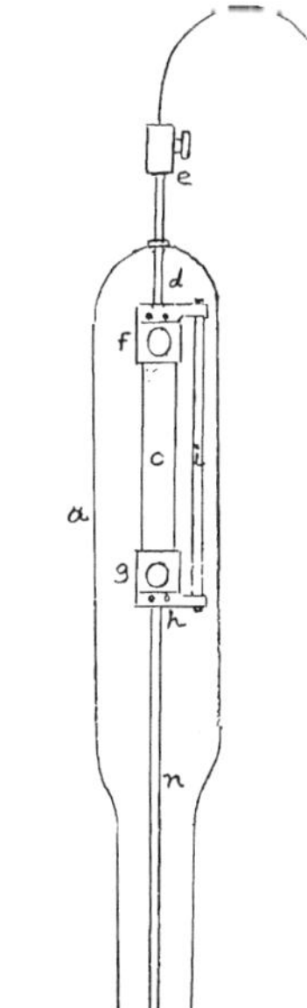

American inventor Hiram Stevens Maxim was one of Edison's fiercest and best-financed lightbulb competitors. He started working on one in 1876. He actually filed a lightbulb patent the day before Edison did, but the courts eventually granted Edison the patent. Maxim left the United States in a huff and moved to Europe, where he invented the modern machine gun in 1883. His weapon was capable of firing more than five hundred bullets per minute, and all modern machine guns are based on its principles. His son, Hiram Percy Maxim, invented the silencer.

DIRTY DISHING

Doing away with dishpan hands

Josephine Garis Cochrane was unhappy with the servants. She didn't think they were handling her fine china carefully enough when washing it. She and her husband often entertained at home, and some of her most valued pieces were getting chipped. So the Shelbyville, Illinois, socialite impulsively decided that she would start washing the china herself.

(No Model.) J. G. COCHRAN. 3 Sheets—Sheet 2.
DISH WASHING MACHINE.
No. 355,139. Patented Dec. 28, 1886.

That didn't last long.

Frustrated by the hateful task, she wandered out of the kitchen and into the study, where she sunk down in a chair to think. There she conceived the idea for a machine that would automatically wash the dishes by spraying hot water on them.

Fired up by her idea, she started working on her dish-washing machine in a shed behind the house. When her husband died shortly thereafter, leaving her in debt, she worked all the harder. Once she had a prototype built, she had to battle with professional mechanics at the manufacturing plant, who dismissed her ideas simply because she was a woman.

But she prevailed.

The Garis-Cochrane Dish-Washing Machine, patented in 1886, became the world's first commercially successful dishwasher. Cochrane thought that it would be a hit with homemakers, but it was hotels and restaurants that proved her best customers. She died in 1912, and it wasn't until more than forty years later that her company—now called KitchenAid—was able to make good on her dream of turning the dishwasher into a common kitchen appliance.

"I WOULDN'T ADVISE ANY WOMAN WHO WANTS TO GET RICH BY HER OWN EFFORTS TO INVENT A DISHWASHER."

—JOSEPHINE COCHRANE

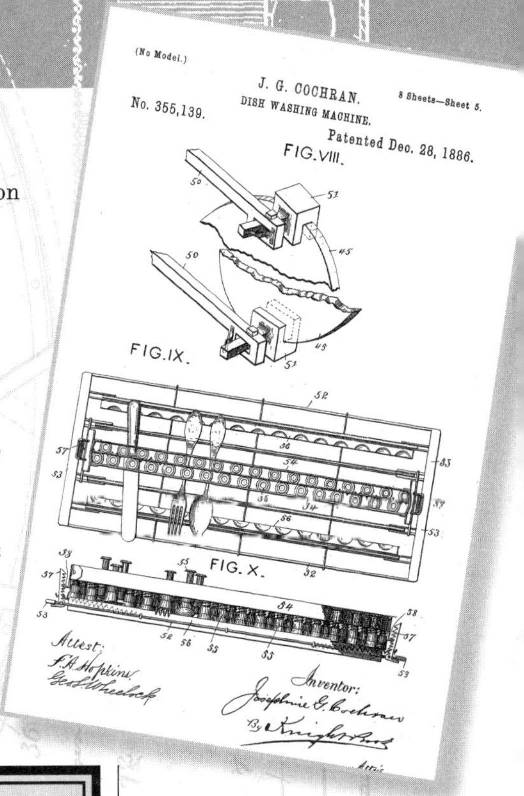

(No Model.)

J. G. COCHRAN. 8 Sheets—Sheet 5.

DISH WASHING MACHINE.

No. 355,139. Patented Dec. 28, 1886.

FIG.VIII.

FIG.IX.

FIG. X.

Attest:

Inventor:

Cochrane's big breakthrough came at Chicago's 1893 Columbia Exposition, which drew millions of visitors. Not only were nine of her machines used at the fair to wash thousands of dirty dishes every day, but her invention was awarded the highest prize by judges evaluating displays in the Machinery Hall.

Cochrane's early machine used wire racks to hold the dishes inside a wheel that spun around inside a copper boiler. The apparatus could either be cranked by hand or driven by a motor.

1890

FADE TO BLACK

An early film mystery

Louis Aimé Augustin Le Prince might have earned everlasting fame as the inventor of the movie camera, if he hadn't disappeared into thin air after getting on board the 2:42 p.m. train to Paris.

Le Prince was a French painter and photographer who began experimenting with movie cameras in the 1880s. By 1886 he had a patent, and by 1888 he had a working camera. In fact, the oldest existing movies in the world, each just a couple of seconds long, were shot by Le Prince that year in Leeds, England. This was well before Thomas Edison or the Lumière brothers had a working a camera.

Le Prince kept making improvements, and by 1890 had worked out all the basics for a system to shoot and project film. Knowing that he was in a race against time with other inventors, he decided to head to New York to exhibit his invention. But first he went back to France to take care of some business and visit his brother in Dijon. On September 16, 1890, he left for Paris, waving good-bye to his brother as the train pulled out of the station,

Le Prince was never seen again. Neither was his luggage.

Had he lived, Le Prince would undoubtedly have continued improving his invention, and might have become the big star in the story of movie camera invention. Instead, Edison introduced his camera in 1893 and won the patent battle that followed. And Le Prince was left on the cutting-room floor.

D OF AND APPARATUS FOR PRODUCING ANIMATED PICTURES
OF NATURAL SCENERY AND LIFE.

76,24 Patented J 88.

(No Model.) 3 Sheets—Sheet 2.
A. LE PRINCE.
METHOD OF AND APPARATUS FOR PRODUCING ANIMATED PICTURES
OF NATURAL SCENERY AND LIFE.
No. 376,247. Patented Jan. 10, 1888.

WITNESSES: INVENTOR: BY ATTORNEYS.

One of the cameras Le Prince developed had sixteen lenses that would each go off in turn, shooting sixteen frames per second. He also developed a single-lens camera with flexible film loaded in spools. Le Prince called his camera a "receiver," and called the projector a "deliverer."

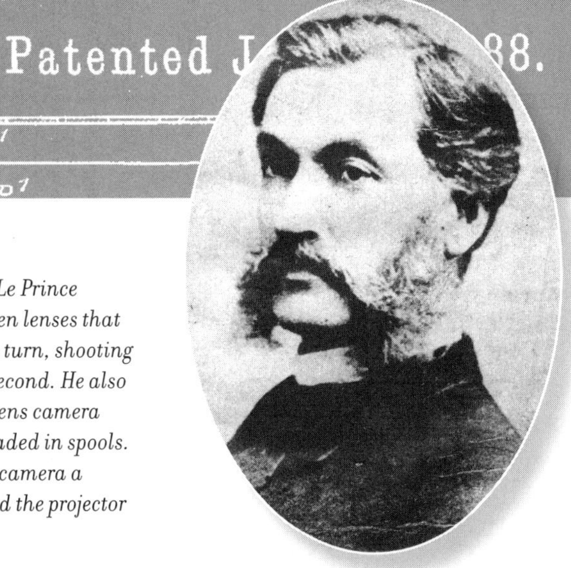

Le Prince's family suspected foul play, but never laid their hands on any proof. Others have theorized that he might have committed suicide. In 2003, an 1890 photo of a drowned man resembling Le Prince turned up in the Paris Police archives. Mysterious tragedy continued to dog the family: two years after testifying in the Edison patent case, Le Prince's son was found shot dead on Long Island.

These are frames from what is believed to be the first real motion-picture ever made. It was shot in October of 1888, in the garden of Le Prince's in-laws, in a town called Roundhay, which is a suburb of Leeds, England.

1891

THE UNDERTAKER'S REVENGE

Saying good-bye to the "hello girl"

Kansas City undertaker Almon Brown Strowger was incensed at the operators manning the local telephone exchange. The telephone had only been invented a decade before, and every call had to be connected by an operator. Strowger was convinced that one of the "hello girls," as the operators were called, was diverting business calls away from him and to a rival undertaker who also happened to be her boyfriend. His indignation made him want to do away with *all* telephone operators.

So he came up with a way to do just that.

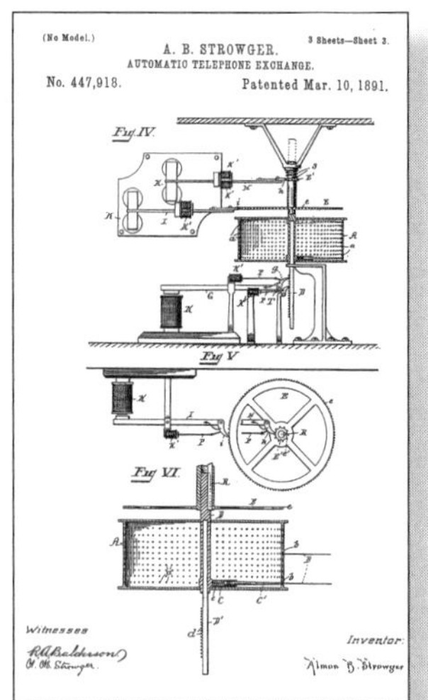

Strowger set out to build an automatic switching device so that people could call him—or anyone—directly. His first model was built with pencils and pins inside a shirt-collar box. By 1891, he had perfected an automatic switch that used electromagnetic impulses to connect calls. Customers used five mechanical buttons to punch in another customer's number, though he quickly got rid of the buttons and introduced the first dial phone.

None of the big phone companies were interested, so he formed a company to market it himself. The first automatic telephone exchange was installed in LaPorte, Indiana, in 1892.

Advertisements called it the "girl-less, cuss-less, out-of-order-less, and wait-less telephone exchange," and it soon went into use across the nation. The job of "hello girls" ceased to exist—rubbed out by a Kansas City undertaker.

Strowger's first automated exchange in LaPorte, Indiana, had fifty-two subscribers. Reporters, investors, politicians, and scientists came from around the world to marvel at a telephone system that didn't require operators.

Strowger's switching system—and dial phones—remained in use for more than seventy years.

Strowger had to assure reporters that he did not hate "hello girls," he just wanted to make telephoning more efficient.

1892

MATCHMAKERS

An invention that was a striking success

Joshua Pusey was a proud veteran of the Civil War, a successful attorney, a prolific inventor, and, above all, a lover of cigars. Pusey often attended fancy banquets in Philadelphia, and he disliked having to stick a bulky box of safety matches in his pocket to feed his cigar habit.

So he set out to invent a "flexible match." With the help of a junior partner named John Raymond Nolan, he snipped some matches out of cardboard with the office shears, and cooked up chemicals on the office's potbellied stove to coat the match heads. Pusey's idea was to insert the matches into a folded cardboard container with a striking surface on the inside.

And so the matchbook was born.

Pusey's patent was eventually purchased by the Diamond Match Company for $4,000. A spirited sales exec named Henry Traute was put in charge of marketing the new product. Traute made two key changes. He moved the striking surface from the inside to the outside to reduce the risk of accidentally lighting all the matches at once, and he coined the familiar phrase that has graced billions and billions of matchbooks since:

"Close Cover Before Striking."

Traute marketed the matchbooks as an advertising medium. Businesses resisted the new idea, but in 1896 his persistent efforts resulted in his first big order: 10 million matchbooks from the Pabst Brewing Company in Milwaukee. The matchbook had finally caught fire.

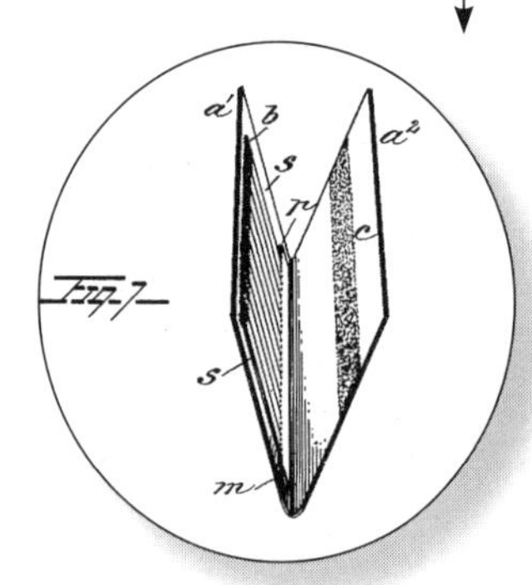

The first advertising on a matchbook appeared in 1895, when the Mendelssohn Opera Company in New York purchased several boxes of blank matchbooks and had cast members decorate them with cutout pieces of photos and hand-scrawled slogans. The only remaining example of their marketing effort is today owned by the Franklin Mint and believed to be worth more than $25,000.

J. Pusey
Nut Cracker.
Patented Nov. 8, 1870.
No. 109,144.

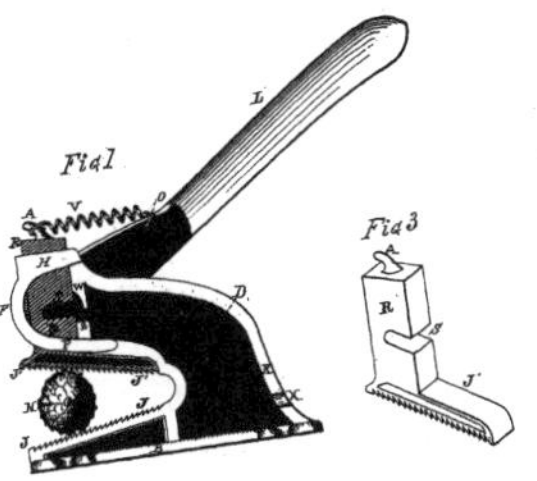

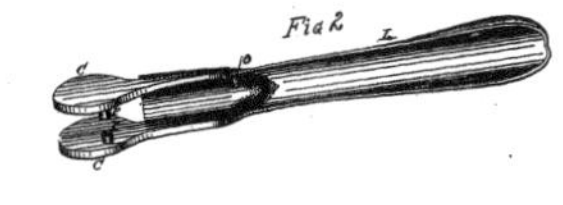

It was amazing that Joshua Pusey had any time to devote to his legal career, given the number of inventions he turned out. The matchbook was the most successful, but it was only one of dozens he patented, including designs for a nutcracker, a cigar cutter, a toboggan, an eraser holder, window fixtures, and many more.

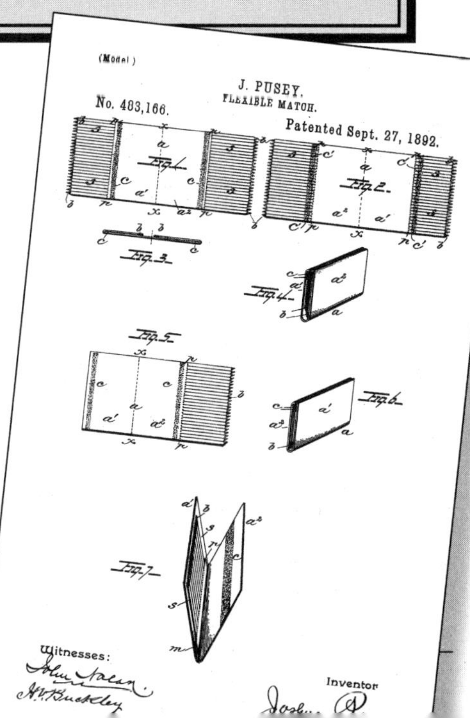
(Model.)
J. PUSEY.
FLEXIBLE MATCH.
No. 483,166.
Patented Sept. 27, 1892.

Witnesses:
Inventor

1893

THE MAN WITH WHEELS IN HIS HEAD

Thinking big in Chicago

Planning for Chicago's Columbian Exposition was well under way, but the man in charge, architect Daniel Burnham, was less than satisfied. He dreamed of an event so amazing it would utterly surpass the recently concluded Paris Exposition. But the Chicago fair had nothing to compare with the spectacular tower designed by Louis Eiffel that wowed visitors to the Paris show.

Burnham put out the call for something better.

Various outlandish towers were suggested, including one two miles high, and another made of logs. But Burnham didn't want another tower, he was looking for something altogether different. That's when an engineer from Pittsburgh had a brainstorm: why not create a giant wheel that could lift passengers hundreds of feet into the sky.

Other engineers said it was so big that it couldn't be built, and the idea was rejected. But George Ferris persevered. Mocked as "the man with wheels in his head," he finally got approval to go ahead. Built in less than six months, his breathtakingly enormous wheel could carry more than two thousand passengers at a time, taking them nearly three hundred feet into the air. More than a million and a half fairgoers rode it into the sky, and the Ferris wheel has been with us ever since.

The first Ferris wheel was also operated at the St. Louis Exposition in 1906. Left standing after the fair, it was considered an eyesore, and was eventually dynamited and sold for scrap. The wheel long outlived its creator, who died in 1896 of tuberculosis.

The wheel weighed more than two million pounds. It carried thirty-six cars, each the size of a railroad car. It took fourteen tons of bolts just to hold everything together.

1893

THE HOOKLESS HOOKER

A modern-day marvel that took decades to catch on!

In 1893, Whitcomb Judson patented a hook-and-eye fastener he called a "clasp locker." It was the world's first zipper. Judson demonstrated his invention at the Chicago World's Fair, but the world failed to take notice. The Post Office ordered zippers for twenty mailbags, but they worked so poorly that the order was never repeated.

In 1913 an engineer named Gideon Sunback came up with a revamped design. He called it the "Hookless Hooker," because instead of using hooks and eyes, it pressed together tiny interconnecting scoops, just like a modern zipper.

The fastener started appearing on money belts, tobacco pouches, and boots, but it was viewed more as a novelty than a necessity. Clothing manufacturers showed little enthusiasm for the device. That began to change in the 1930s, when zippers started appearing on children's clothes.

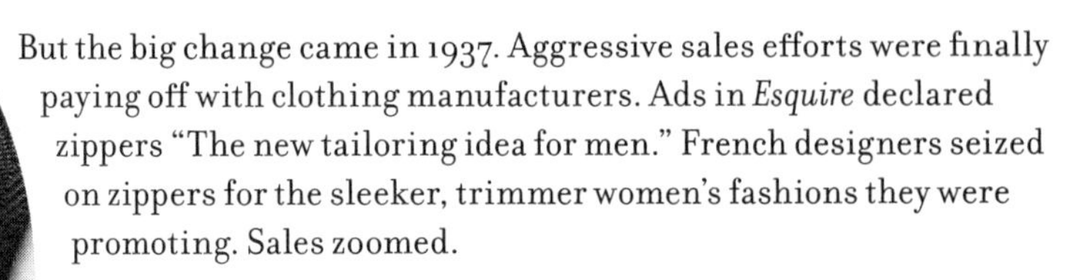

But the big change came in 1937. Aggressive sales efforts were finally paying off with clothing manufacturers. Ads in *Esquire* declared zippers "The new tailoring idea for men." French designers seized on zippers for the sleeker, trimmer women's fashions they were promoting. Sales zoomed.

Forty-four years after its invention, the zipper was finally unstuck.

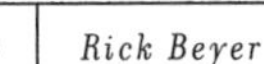

Elsa Schiaparelli, who coined the term "shocking pink," was the first of the Paris fashion designers to embrace zippers for women's clothing. Her pioneering designs of the 1930s used bright-colored zippers on ladies gowns.

Goodrich
ZIPPER

Nothing to Button, Hook, Lace or Tie

The Zipper Boot is a marvel of comfort and smartness. It is worn right over your shoes or slippers.

On and off in a jiffy—nothing to button, hook, lace or tie.

The Hookless Fastener—exclusive on Zipper footwear—does the trick. A little pull of the tab and ZIP! · · · it opens wide or locks snug and tight.

Ask your dealer for Zippers. Sizes for men, women and the kiddies.

THE B. F. GOODRICH RUBBER COMPANY
Akron, Ohio ESTABLISHED 1870

In 1923, Goodyear brought out a new line of boots with the fasteners and called them "Zipper Boots," because of the sound the fastener made when you closed it. The name stuck.

1895

CUTTING EDGE

He set out to revolutionize society. He succeeded.

King Camp Gillette was a salesman, a tinkerer, and a dreamer. While traveling the country peddling products, he dreamed of creating a utopian society free of poverty, crime, and war. He also dreamed of inventing his way to fame and fortune, but couldn't cash in on any of his patents.

What changed his life was the bottle cap.

Gillette worked as a salesman for bottle-cap inventor William Painter. He told Gillette that the key to success was coming up with a product that customers would keep throwing away and buying more of. Gillette spent years making lists of possible disposable products.

One morning in 1895, he picked up his razor to shave. Razors in those days had thick steel blades that required frequent stropping to keep them sharp. Gillette's razor was so dull that stropping wasn't going to be enough—he needed to take it to a cutler to have it honed.

And then it hit him.

Gillette sold fifty-one razors and less than two hundred blades the first year his product was on the market. The next year he sold ninety thousand razors and fifteen million blades. By the end of his life, Gillette was selling more than twenty million razors a year.

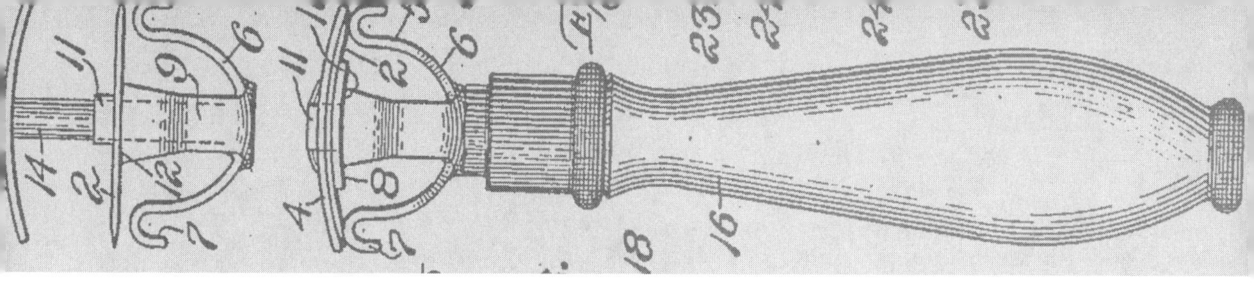

Why not make a paper-thin razor blade so cheap that it could be used and thrown away? It took him eight years to perfect the idea, but the Gillette razor that came to market in 1903 changed shaving forever. It also paved the way to our modern disposable culture—a revolution of sorts, though not exactly the one Gillette had in mind.

> **"I HAVE GOT IT. OUR FORTUNE IS MADE."**
>
> —GILLETTE, IN A LETTER TO HIS WIFE AFTER DREAMING UP THE DISPOSABLE RAZOR

Gillette set out his vision for society in a book called The Human Drift. *His plan called for a giant corporation to acquire all the assets of the world, thus eliminating evils of competition. A utopian city of 60 million would be built near Niagara Falls, whose waters would be harnessed for electrical power. He had more luck selling razors than reform.*

1895

RÖNTGEN'S RAYS

The man who uncovered his own skeleton

Wilhelm Röntgen was expelled from high school and failed his college entrance exam. Yet this slow start to his academic career didn't prevent him from winning a Nobel Prize—for something he discovered by accident.

On November 8, 1895, Röntgen was experimenting with a new type of cathode ray tube to better understand its properties. He was still setting things up in his darkened lab when he noticed that the tube was causing a specially coated screen across the room on his workbench to give off a strange glimmering light. It couldn't be the cathode rays themselves—they only traveled a few inches.

Something else must be coming out of the tube. Something no one else had noticed.

Excited, Röntgen kept experimenting. He discovered that the rays emitted by the tube could be blocked by lead. One day, while placing a piece of lead in front of the tube, he saw the shadow cast by his hand—and was shocked that he could see the bones inside of it.

In mathematics, *x* signifies the unknown. So it was only natural that Röntgen used that symbol to provide a name for the mysterious rays he had discovered:

X-rays.

It was proposed that the rays be called "Röntgen's rays," but he opposed the idea. He did, however, accept the first ever Nobel Prize for Physics, awarded in 1901.

Few discoveries have created a bigger sensation than this one. Within days of the announcement, it was making headlines around the world. "All hell broke loose," said Röntgen. "In a few days I was disgusted with the whole business." A London company advertised X-ray-proof underwear. A French scientist announced that he had used X-rays to photograph the soul.

Röntgen took the first X-ray photo of his wife's hand. Anna Bertha Röntgen felt as if she was seeing a ghost of her hand, and refused to have it ever done again. Within six weeks of the discovery, a doctor in Berlin performed the first surgical operation done with the help of X-rays—the removal of gunshot fragments embedded in a young man's hand.

PATENTLY ABSURD I

Seemed like a good idea at the time

Not every invention is a winner. And some are downright bizarre. Here are a few of the stranger items for which the U.S. Patent Office has seen fit to grant a patent.

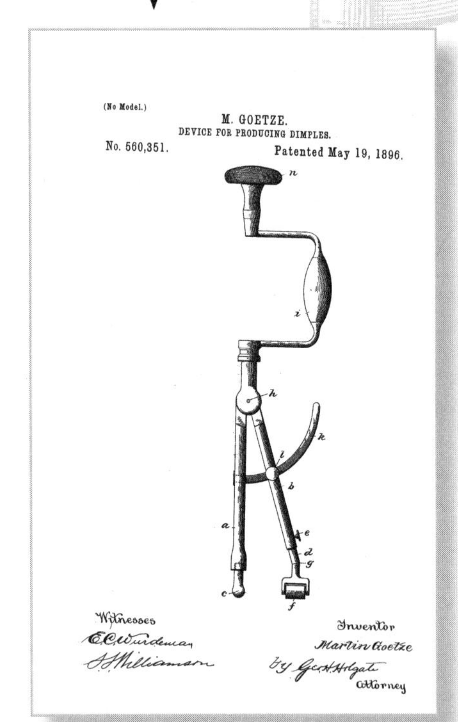

Martin Goetze's nifty little 1896 invention promised to "produce dimples" or to "nurture and maintain" pre-existing dimples, although it is difficult to imagine said dimpled person smiling at the memory.

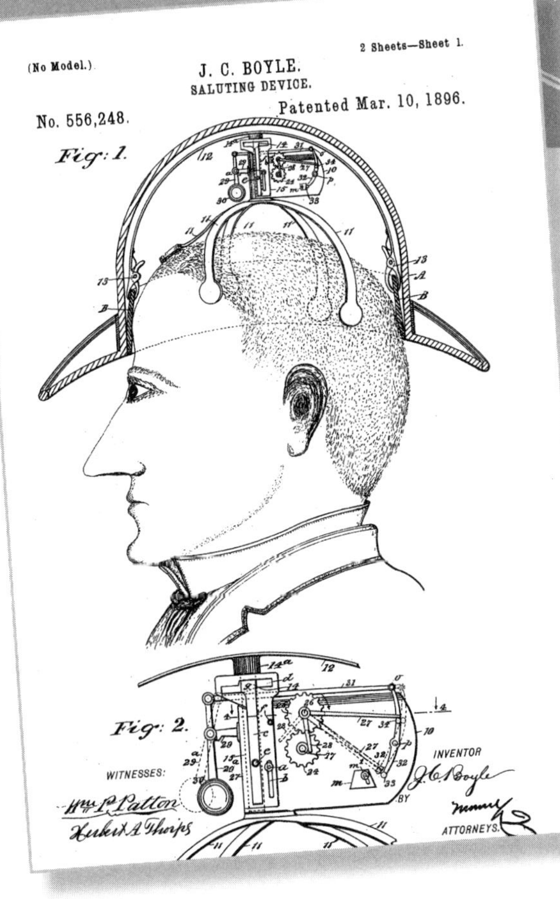

J. C. Boyle's 1896 invention enabled a gentleman to "effect a polite salutation" by tipping his hat, even when his hands were full. Simply by nodding, he activated weights and gears designed to lift the hat, rotate it, and bring it gracefully back down for a landing. Comfort does not seem to have been an issue.

Escape from burning buildings is something that many inventors have contemplated, perhaps while traveling. A Tennessean named Benjamin Oppenheimer thought he had licked it in 1879 with a parachute attached to the head paired with cushioned boots for a smooth landing. Hungarian-born inventor Michael Kispéter came up with an improved version in 1915, featuring an arm-activated parachute, and in case things went really *badly, a spring-loaded hat, presumably so that a person landing upside down would bounce gently to safety. How optimistic!*

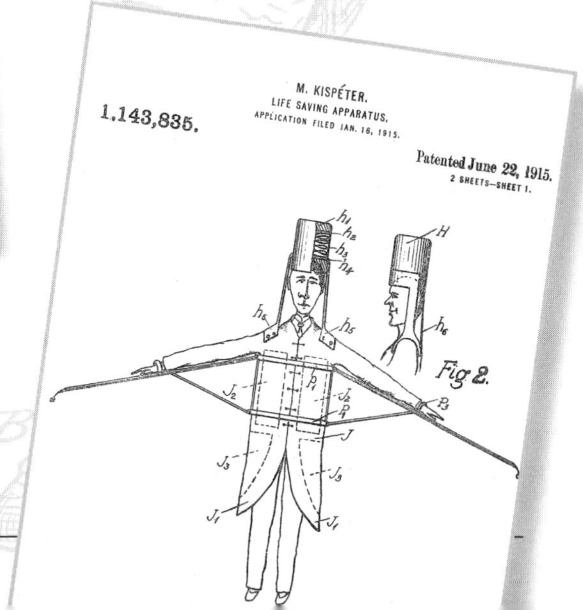

THAT GIANT SUCKING SOUND

The men behind the machines that conquered dirt

In 1901, British engineer H. Cecil Booth saw a demonstration of a new type of cleaning machine. But it suffered from an obvious flaw, he thought: it just blew dust and dirt around. He asked the inventor why he didn't create a machine that simply sucked the dirt up. Not practical, he was told.

Booth thought otherwise.

To prove to himself that suction would work, he took a handkerchief and put it over his mouth. Then he leaned into a dirty chair and took a deep breath. Aha! A black ring appeared on his improvised filter. Booth promptly went to work on a suction machine.

Later that year he patented the world's first vacuum cleaner. It was a massive machine mounted on a horse-drawn wagon. Booth named it a "Puffing Billy" after a famous early locomotive. The cleaner would stay on the street while a hose was snaked up the side of a building and fed through a window. It became wildly popular.

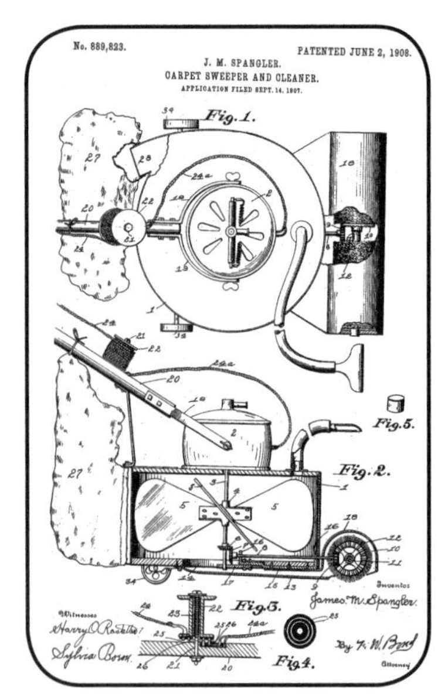

Booth's vacuum cleaner proved its worth during World War I, when a spotted-fever epidemic was killing soldiers billeted at London's famous Crystal Palace. A fleet of Puffing Billies sucked up an astonishing twenty-six tons of dirt from the giant building's carpets . . . and the spread of the disease was halted virtually overnight.

In the early days of vacuum cleaners, blue-blooded ladies in London would often hold tea parties during cleanings to show off how up to date they were.

James Murray Spangler was a fifty-nine-year-old janitor in Canton, Ohio. He suffered from asthma, and sweeping the floors every night left him gasping for breath. In 1907, he fashioned a homemade machine from a soapbox, a pillowcase, and a broom handle. An electric motor powered a brush and fan that stirred up the dirt and blew it into the pillowcase. He called it a suction sweeper. It worked so well that he started making ones for friends and relatives. One happy buyer was a woman whose husband owned a leather goods company. His name was H. W. Hoover. He bought the rights from Spangler in 1908 and the first Hoover vacuum cleaner hit the market three months later.

1903

ACCIDENTS HAPPEN

The laboratory mishap that helped make cars safer

When French scientist Edouard Benedictus knocked a beaker onto the floor, he expected it to shatter—and it did. But what puzzled him was that the pieces didn't break apart. The shattered beaker still held its form.

Upon investigating, he discovered that the beaker had contained a solution of nitrocellulose, a form of liquid plastic. The solution had evaporated, leaving a thin coat on the inside. That's what held the beaker together even after the glass had cracked every which way.

Benedictus didn't think much about the incident until a few years later, when he heard about a young girl whose face was badly disfigured by broken glass in a car accident.

"Suddenly there appeared before my eyes an image of the broken flask. I leapt up, dashed to my laboratory, and concentrated on the practical possibilities of my idea." Benedictus worked twenty-four hours straight in his lab to laminate two pieces of glass together with a coat of celluloid adhesive between them.

Laminated safety-glass would go on to uses in thousands of items, from automobiles to eyeglasses.

And it all started with the beaker that didn't break.

Carmakers started using safety glass in the 1920s. Before that, an estimated 50 percent of all persons hurt in car accidents were cut by pieces of broken glass.

1903

A WIRE IN WINTER

The all-but-forgotten inventor with a twisted hang-up

On a cold November day in 1903, Albert Parkhouse returned from lunch to his job at the Timberlake Wire and Novelty Company in Jackson, Michigan. He discovered, much to his chagrin, that there was no place to hang his winter coat. All the hooks were taken.

Not wanting to throw his good winter coat over a chair and risk it getting rumpled, Parkhouse picked up a piece of wire—after all, he worked at a wire company—and twisted it into two loops that would fit inside his coat. Then he twisted the ends into a hook he could hang up some place.

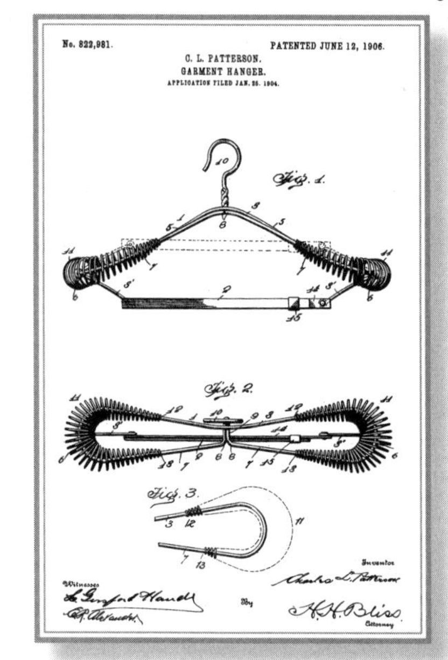

The modern wire coat hanger was born.

Other employees started to copy the idea, and his employer decided to patent it. Patterson earned nothing from his ingenious invention, not even the honor of having his name on the patent. The company's attorney put *his* name on it instead!

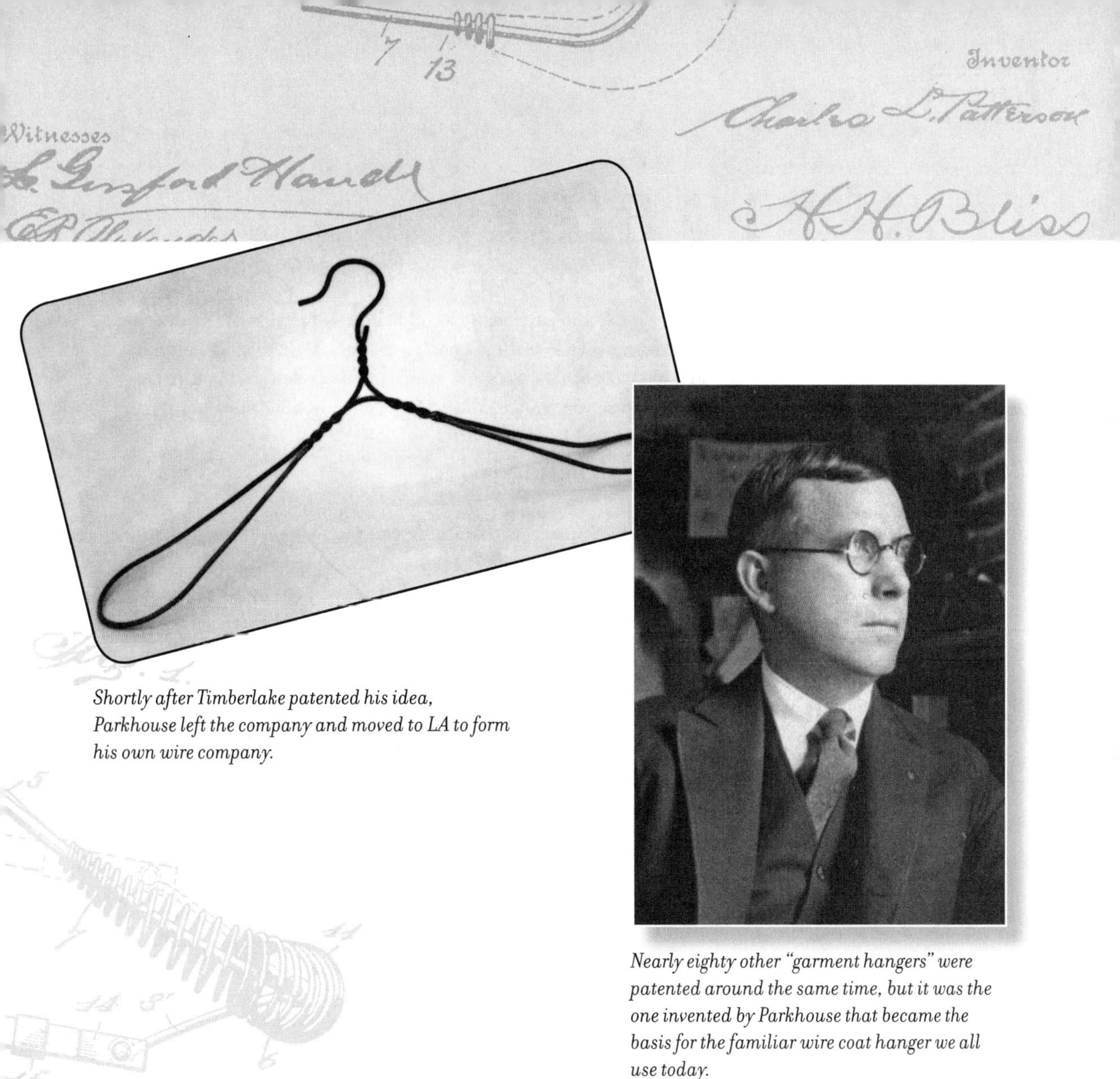

Shortly after Timberlake patented his idea, Parkhouse left the company and moved to LA to form his own wire company.

Nearly eighty other "garment hangers" were patented around the same time, but it was the one invented by Parkhouse that became the basis for the familiar wire coat hanger we all use today.

1905

TICKET TO RIDE

A commute for the ages

In May of 1905, a twenty-six-year-old clerk in Bern, Switzerland, was headed home on the trolley. Deep in thought, depressed about his inability to solve a nagging question that had been vexing him for years, he glanced up at one of the city's most famous sites: its ornate clock-tower.

In that instant, the world changed.

The clerk was named Albert Einstein. He was puzzling over how to square new theories regarding the speed of light with the old laws of physics laid out by Newton. The new theories suggested the speed of light was constant—no matter what. But that created all sorts of paradoxes that he couldn't resolve.

As he glanced at the clock, he wondered how it would look if his trolley was traveling at the speed of light. It would look frozen, since the light could not catch up to the streetcar. But a clock sitting in the seat next to him would tick away normally.

Then, suddenly, in Einstein's words, "A storm broke loose in my mind."

The clock provided the key. The speed of light was constant—but *what if time itself was not*? What if time could unfold at different speeds, depending on how fast an observer was traveling? Then everything would fit together.

This remarkable revelation led Einstein to his theory of

wo man zuweilen aus scheinbar ganz fernliegenden Thatsachen
noch nicht beobachtete Erscheinungen schliessen kann
einem solchen Gefühl von Sicherheit, dass man ohne Herzklopfen
sogar ohne Neugier den Vergleich mit der Erfahrung erwartet.

Albert Einstein

special relativity, and the formula E=mc², which in turn helped lead to a new understanding of the universe and the development of nuclear power.

Worth the price of a trolley ticket.

Bern's clock tower, known as the Zytglogge ("time bell"), was built in the 1200s as part of the medieval city's west gate. The clock itself dates from the 1400s. The famous tower dominates the Swiss city's main square.

Einstein spent six weeks working out the mathematics of his theory. Then, according to his son, he went straight to bed for two weeks—but not before giving the paper to his wife Mileva, a talented mathematician in her own right, to check for errors. She was so confident his work would win him a Nobel Prize that when they were divorced in 1919, she demanded that he turn over to her any future Nobel Prize award. Einstein forked over the dough when he was awarded the Nobel in 1921.

1910

SWEET CARESSE

The shocking socialite determined to defy convention

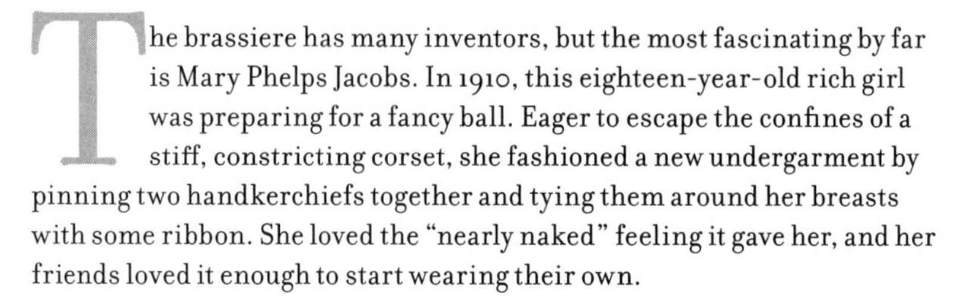

The brassiere has many inventors, but the most fascinating by far is Mary Phelps Jacobs. In 1910, this eighteen-year-old rich girl was preparing for a fancy ball. Eager to escape the confines of a stiff, constricting corset, she fashioned a new undergarment by pinning two handkerchiefs together and tying them around her breasts with some ribbon. She loved the "nearly naked" feeling it gave her, and her friends loved it enough to start wearing their own.

Jacobs went on to become the first person to obtain a U.S. patent for a brassiere. But that was only the beginning of life dedicated to doing things differently.

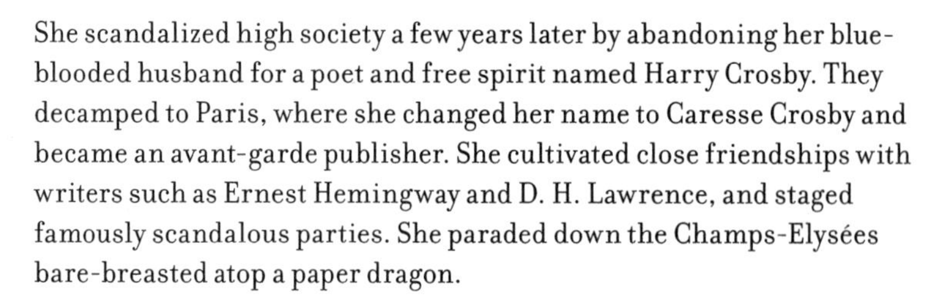

She scandalized high society a few years later by abandoning her blue-blooded husband for a poet and free spirit named Harry Crosby. They decamped to Paris, where she changed her name to Caresse Crosby and became an avant-garde publisher. She cultivated close friendships with writers such as Ernest Hemingway and D. H. Lawrence, and staged famously scandalous parties. She paraded down the Champs-Elysées bare-breasted atop a paper dragon.

After her second husband and his lover killed themselves in a sensational suicide pact, Caresse collected two more husbands and numerous lovers, including a Russian prince, an English earl, and an African-American boxer-turned-bandleader-turned-actor named Canada Lee. In the 1940s she opened the first modern art gallery in Washington, DC, and introduced artist Salvador Dali to America. At the end of her life, she was trying to establish a center for world peace in an Italian castle.

She died in 1970, after seventy-eight years of refusing to follow the rules.

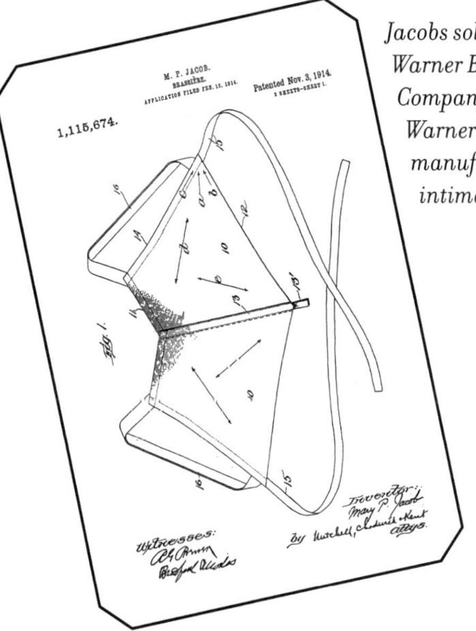

Jacobs sold her patent to the Warner Brothers Corset Company for $1,500. Warner's continues as a manufacturer of women's intimate apparel today.

Corsets, braced with metal and whalebone, were the fashion for more than a hundred years. They were not only extremely uncomfortable but were worn so tightly (to create an hourglass figure) that they proved dangerous to a woman's internal organs.

The brassiere didn't really take off until a young seamstress who had emigrated from Minsk to Manhattan designed her own in the 1920s. Feeling that her dresses didn't look as good on her clients as they should, she created a bra that lifted and shaped a woman's breasts instead of hiding them. "Nature has made woman with a bosom," said Ida Rosenthal. "Who am I to argue with nature?" Eventually, she started a new company: Maidenform.

1911

FOREVER YOUNG

The tinkerer who invented a better way to tinker

By the time he was twenty-seven, Alfred Gilbert had some notable accomplishments to his name. He had won an Olympic gold medal for pole vault in the 1908 games. He also had earned a medical degree from Yale. And he was an accomplished magician who had started a company that sold kits for other budding magicians.

But inside the successful young man was a boy who refused to grow up. And that was to lead to a success that would dwarf all the others.

In 1911, he was riding the New York, New Haven, and Hartford Railroad into New York City. The line was being converted from steam to electricity, and Gilbert was fascinated by the girders erected to carry the power lines. It would be fun, he mused, to have miniature girders like that to play with at home.

That night, at the dining table, he cut some miniature girders out of cardboard. He took them to a toolmaker, and asked him to make them in steel. Soon he designed other parts and a miniature motor so he could build almost anything.

The result was a toy that came out in 1913 and was to thrill the hearts of budding engineers for more than seventy-five years. Gilbert called it "Erector, the World's Greatest Construction Toy for Boys." But the boys themselves had a simpler name for it.

The Erector Set.

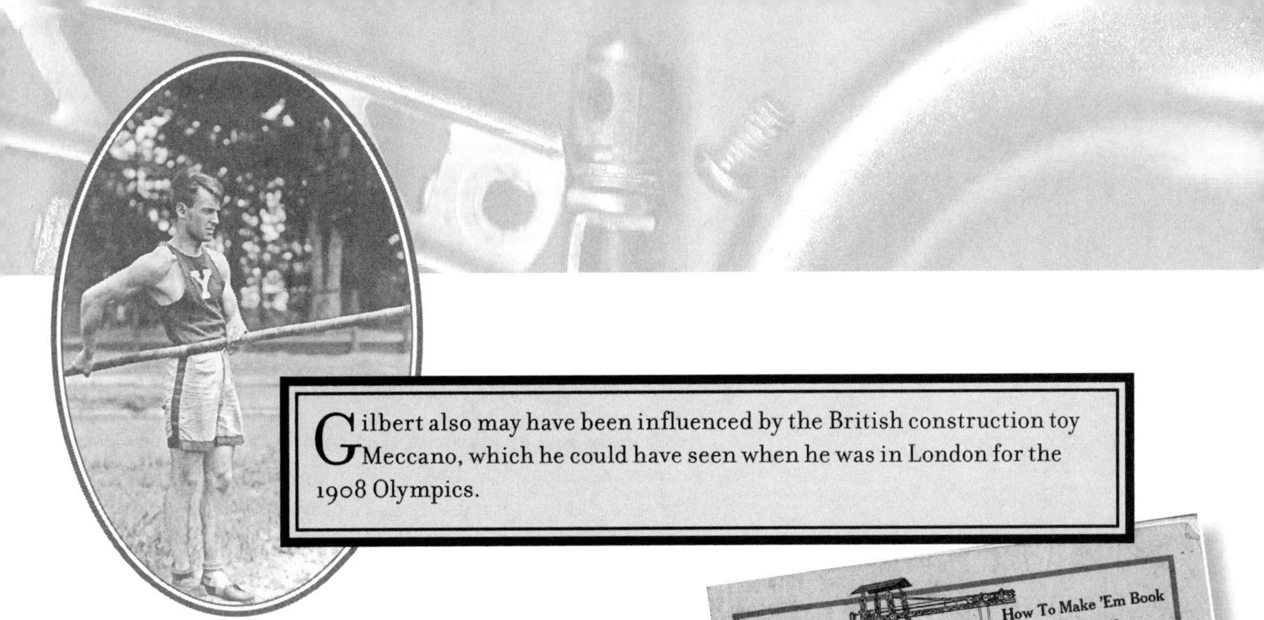

Gilbert also may have been influenced by the British construction toy Meccano, which he could have seen when he was in London for the 1908 Olympics.

In 1918, the government's council on national defense was considering a ban on Christmas toy sales to help the war effort. Gilbert led a delegation of toy manufacturers to plead their case, making sure they brought along a selection of their latest wares. The august members of the council (including Secretary of War Newton Baker) were soon playing enthusiastically with the toys, and the idea of the ban was dropped. Gilbert became nationally known as "The Man Who Saved Christmas."

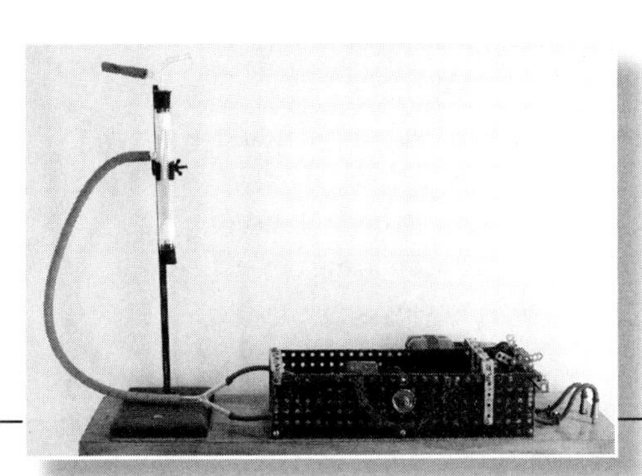

In 1949, Yale medical student William Sewell created one of the world's first artificial hearts, out of pieces from his Erector Set and $20 worth of other spare parts. It was powered by the Erector Set's motor. The external pump was tested on a dog, and successfully bypassed the animal's heart for more than an hour. (The dog made a full recovery.) When Dr. Michael DeBakey pioneered heart bypass surgery in the early 1960s, he used a pump nearly identical to Sewell's.

1912

THE AMAZING DOCTOR ABRAMS

A medical miracle that seemed too good to be true

Dr. Albert Abrams revolutionized American medicine in 1912 with a new way of looking at disease. "The spirit of the age is radio," he said, "and we can use radio in diagnosis." Abrams developed a breakthrough technique known as spancho-diagnosis, which used an electronic device called a dynamizer to analyze patients' blood samples.

Abrams's radical theory was that different diseases gave off different "electronic reactions." According to Abrams, the miraculous dynamizer could not only diagnose disease but determine patients' age, race, and even their religion.

His medical technology was seen as a lifesaver. Many people who thought themselves healthy were diagnosed as having traces of such diseases as syphilis, cancer, or malaria. Luckily, Abrams also discovered a cure. He invented the oscillocast, which could be tuned to create vibrations at the same frequency as the disease, attacking it the same way an opera singer's high note can shatter a glass.

Critics scoffed, but Abrams was lionized by a grateful nation. He lectured to packed crowds across the country. He leased out thousands of his machines, and made millions from his success. To protect the technology, he forbade anyone from opening one of his devices and looking inside. The prolific doctor also pioneered

spondylotherapy and invented the electro-concussor and the reflexo-phone.

Have you figured out Dr. Abrams secret? He was a con man, a charlatan, a phony. He pulled off the greatest medical hoax in history, and went to his grave a rich man, just as the truth was spilling out.

Abrams generated lots of media coverage and won over millions of people, but attacks mounted on him in the last years of his life. Nobel Prize-winning physicist Robert Millikan examined one of Abrams's devices as part of a court case in 1923. "It's a contraption which might have been thrown together by a ten-year-old boy who knows a little about electricity to mystify an eight-year-old boy who knows nothing about it."

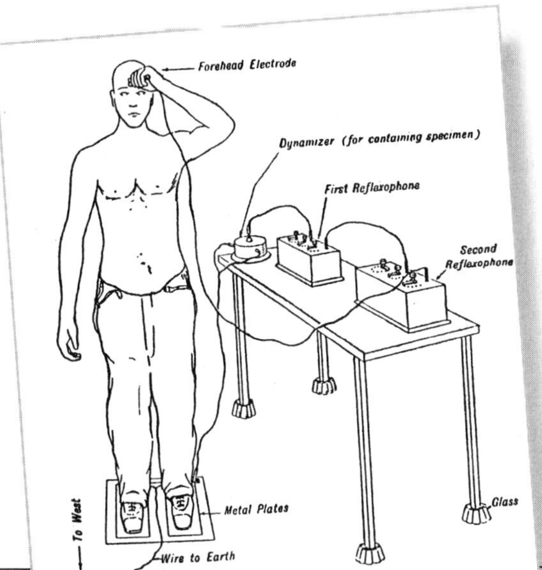

Some of the details of how Abrams's system worked should have been enough to give people pause. Patients had to have their blood drawn in a darkened room, facing west, with their arms raised. Abrams also claimed that the machine could work with a sample of the person's handwriting instead of their blood!

1914

TURNING POINT

The world's first movie star was also an automotive pioneer

Florence Lawrence started appearing in films in 1907. She made dozens of short films for the Biograph production company, and struck a chord with a public that was wild about the new film industry. In those days, movies didn't have credits, and audiences didn't know the names of the people appearing in them. So she was known only as "The Biograph Girl."

But in 1910 she switched studios. The head of the new studio, Carl Laemmle, paid her the unheard-of sum of $1,000 dollars a week. He also promoted her by name through publicity stunts and advertisements.

In other words, he made her the world's first movie star.

Florence used some of her newfound wealth to buy one car, and then another and another. She became an enthusiastic driver. "A car to me is something that is almost human," she said, "something that responds to kindness and understanding and care, just as people do." Lawrence didn't just drive her cars, she set out to make them better. By 1914, she had invented the first electric turn signal. Lawrence's "auto signaling arm" was mounted on the back fender. The driver could push a button, and raise an arm pointing in the direction of the turn. She also invented a brake signal that raised a sign saying "STOP" when her foot hit the brake.

Although the mechanically inclined movie maven never took out a patent, she was way ahead of her time. Electric turn signals didn't become standard on cars until 1939, a year after her death.

Lawrence's film career declined after 1920, and by the time she committed suicide in 1938, the one-time star was all but forgotten.

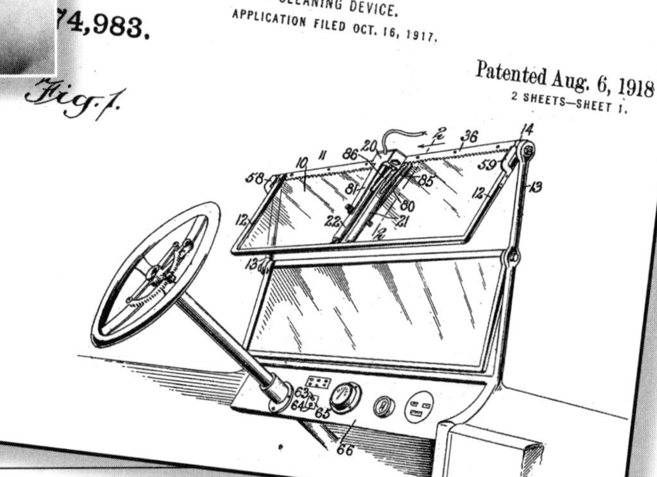

Lawrence's mother, Charlotte Bridgwood, was also an automobile inventor of some note. She patented an early powered windshield wiper in 1917. Mother and daughter formed a company to sell the device, but it was not a commercial success. Another female inventor, Mary Anderson, patented the very first windshield wiper, operated by hand, back in 1903.

1920

JOSEPHINE'S JOY

Clumsiness in the kitchen inspires first-aid fixture

In 1917, Earle Dickson married Josephine Frances Knight. But their marital bliss was marred by minor mishaps Josephine suffered in the kitchen. She always seemed to be cutting or scraping herself.

Earle sought a convenient way to deal with Josephine's numerous nicks. He laid out a piece of surgical tape and stuck little folded-up pieces of gauze at intervals on it. He put a strip of crinoline—a stiff material often used in petticoats—over the whole assembly and rolled it back up again. The result was a prefab bandaging system ready for action. Whenever Josephine cut herself, she could just cut a piece off and fix herself up.

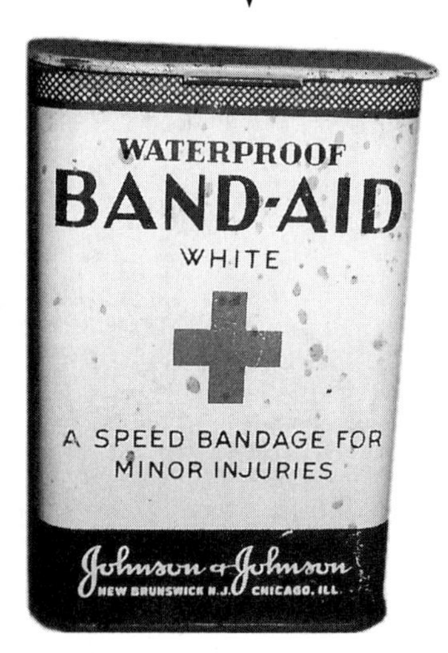

Dickson was a cotton buyer for Johnson & Johnson, and fellow employees suggested he tell management about his idea. He did so, and so (with some tinkering) was born one of the company's most popular products:

The Band-Aid.

Introduced in 1920, it was not an immediate hit. Only three thousand were sold the first year, perhaps in part because it came in a long, 3"-wide roll—people were expected to cut off the part they wanted to use. Johnson & Johnson made the Band-Aids smaller and hit on the idea of distributing them to Boy Scout troops free of charge as a publicity stunt. That did the trick, and people have been stuck on Band-Aids ever since.

It went on the market in 1920, and more than 100 billion have been sold since.

Earle Dickson went on to become a VP at Johnson & Johnson, and later a member of the board of directors. Josephine's clumsiness proved to be the best thing that ever happened to the couple.

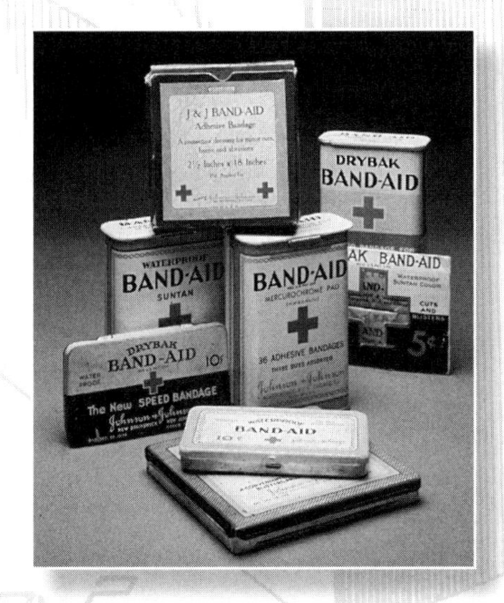

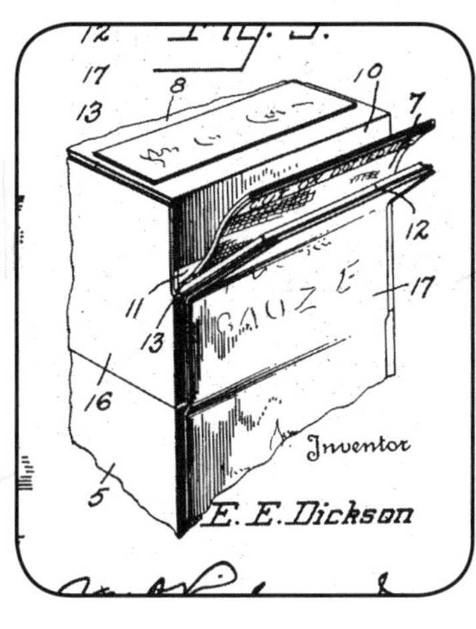

1921

OTTO'S DREAM

Inspiration strikes at 3 a.m.

Austrian scientist Otto Loewi awoke one night from a dream in which he conceived an experiment, simple but brilliant, that would resolve a fundamental question in neuroscience—whether chemicals were involved in the transmission of nerve impulses. He made a few notes on the idea and went back to sleep. When he awoke, he recalled that he had thought of something important, but he couldn't remember what it was. So he looked at his notes. That's when the dream turned to a nightmare.

He couldn't decipher a word of his hasty middle-of-the-night scrawl.

Distraught doesn't begin to describe how Loewi felt. He spent hours trying to remember what he had dreamed. He later called it "the most desperate day in my scientific life."

The next night, the dream recurred. Awaking once again in the middle of the night, Loewi took no chances. He put on his clothes and went straight to his lab. By the time his fellow researchers came in, the experiment was finished. They were so amazed by what he had done that one predicted it would earn him a Nobel Prize.

Good guess.

Loewi's simple experiment, born in the nether world between wakefulness and sleep, demonstrated that nerve impulses are

Loewi had planned to study fine art until his father convinced him he needed to do something more practical.

indeed transmitted chemically. It led him to a decade of experiments that opened the door to new branches of neuroscience, and eventually led to the invention of mood-altering drugs. And in 1936 he was, indeed, awarded the Nobel.

A dream come true.

When the Germans took over Austria in 1938, they promptly arrested Loewi, who was Jewish. He was allowed to leave the country, but only after handing over to the Nazis the money from his Nobel Prize award. He later moved to the United States.

Loewi's experiment involved stimulating a nerve of one frog so that its heart slowed, then taking some of the fluid that the nerve secreted and injecting it into the heart of a second frog. The heart of the second frog also slowed down, demonstrating that the nerve impulse was transmitted by a chemical in the secretion.

1921

THE TEEN WHO INVENTED TELEVISION

And how a field of potatoes inspired his key breakthrough

As a teenager, Philo Farnsworth spent a lot of time thinking about TV. Sounds like a lot of other teenagers. Of course, in Philo's case, television was still a thing of the future—but he had a few ideas about how to change that.

Farnsworth was raised on a farm in Idaho. He was so smart that when he was a freshman in high school, he didn't study math, he taught it. Farnsworth was fascinated by the idea of transmitting pictures through the air. He knew that the key was finding a way to separate an image into component parts that could be converted into electrical current, and then reassembled at the other end.

In the summer of 1921, he plowed his father's potato field, one row at a time, then surveyed the furrows he had plowed. That's when his "Eureka!" moment came: what if he could trap light in an empty jar and transmit it one line at a time on a magnetically deflected beam of electrons? The beam could move back and forth just like his plow in the field.

He sketched the idea on a scrap of paper for his chemistry teacher, who encouraged him to pursue it.

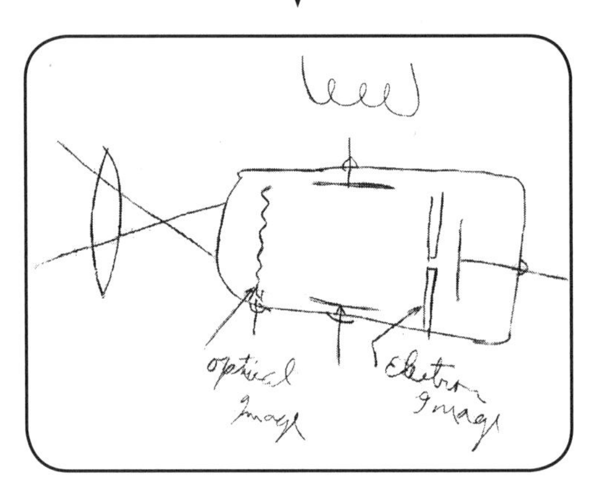

Farnsworth's first sketch, done for his chemistry teacher, Justin Tolman.

Eventually, he dropped out of school and headed to San Francisco, where he set up a lab at the tender age of nineteen to perfect his invention.

Philo Farnsworth dreamed up the idea of electronic television in a farm field at age fifteen. But don't be too impressed with that. After all, he didn't transmit his first picture and patent the idea until he was nearly twenty-one!

Aug. 8, 1939.

P. T. FARNSWORTH

TELEVISION METHOD

Original Filed Jan. 9, 1928

2,168,768

Fig. 1.

Fig. 2.

Fig. 3.

AMPLIFIER

OSCILLATOR

INVENTOR,
PHILO T. FARNSWORTH
BY
ATTORNEY

Farnsworth wouldn't let his son, Kent, watch television, and he himself only appeared on television once. In 1957 he was a guest on *I've Got a Secret*, where he won $80 and a carton of cigarettes after stumping the panel.

The "image dissector tube" was the heart of Farnsworth's all-electronic television. Before it was invented, scientists used a spinning disk with holes in it to mechanically separate a TV picture into pieces. Farnsworth understood that it would be difficult to get the disk to spin fast enough to create a clear picture. He was the first to invent an all-electronic means of making television work, one that is still used today.

1922

TEE TIME

The rich history of the humble golf tee

Traditionally, golfers preparing to tee off would do so from a sandbox known as the tee box. They would grab a handful of sand and shape a little cone from which to hit their ball. Then they would wipe the sand off their hands with a small towel and step up to take their shot. By the late 1800s, this had lost its appeal to some people, and a series of inventors patented different kinds of golf tees. Some lay on top of the ground; others stuck into it. They featured pins and bits of rubber and all sorts of innovations.

All of them had something in common. They were largely ignored by the golfing public. While there were those who didn't want to get their hands dirty, most golfers couldn't imagine the game without sand tees. It was, after all, tradition!

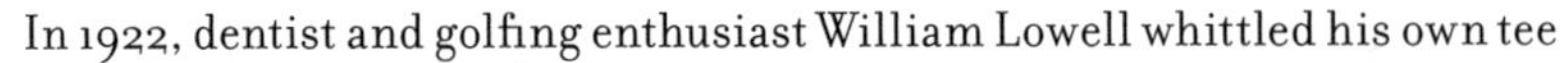

In 1922, dentist and golfing enthusiast William Lowell whittled his own tee using dental tools. He painted it red and called it the Reddy Tee. The shrewd Dr. Lowell also invented something else that helped convince people to pay money for his product.

The sports endorsement.

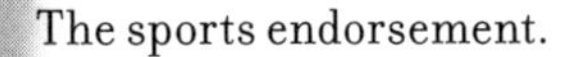

Lowell paid U.S. Open Champ Walter Hagen $2,500 to endorse his product. Suddenly every golfer wanted to use one. Lowell sold more than 70 million Reddy Tees. The tee, and the sports endorsement, have been with us ever since.

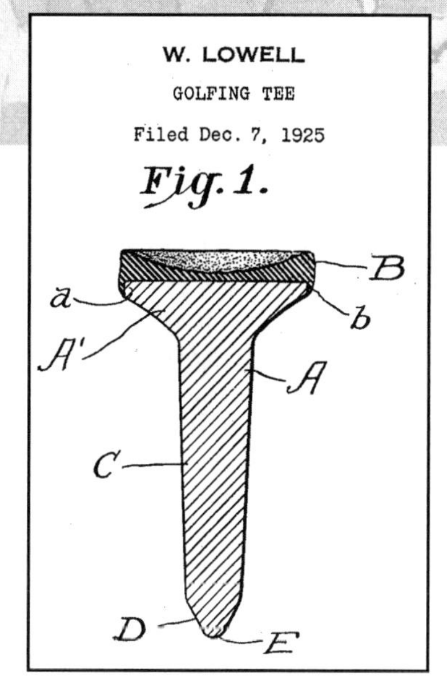

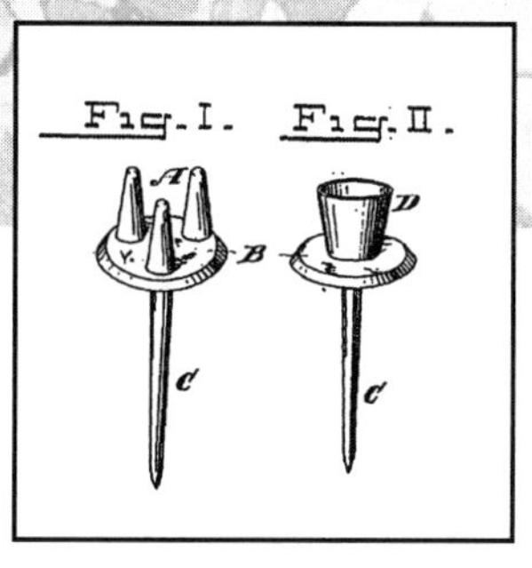

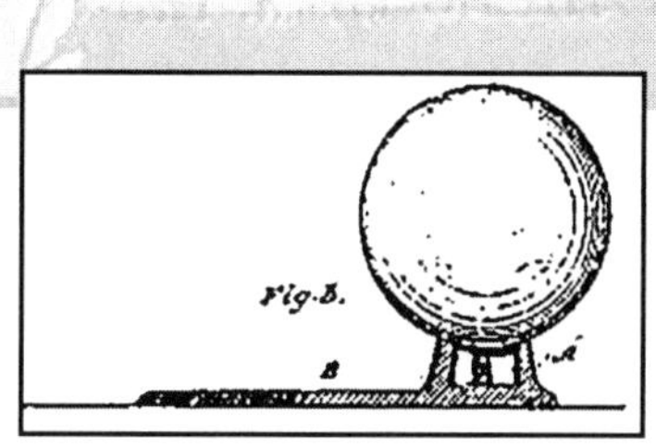

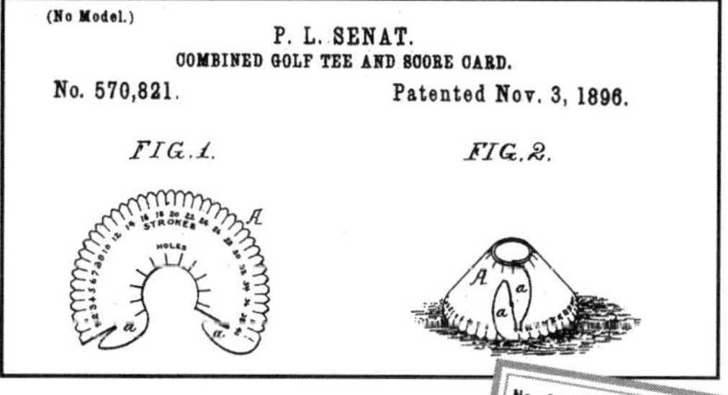

Lowell's patent, along with a few of the tees that were invented earlier, failed to capture the golf world's fancy.

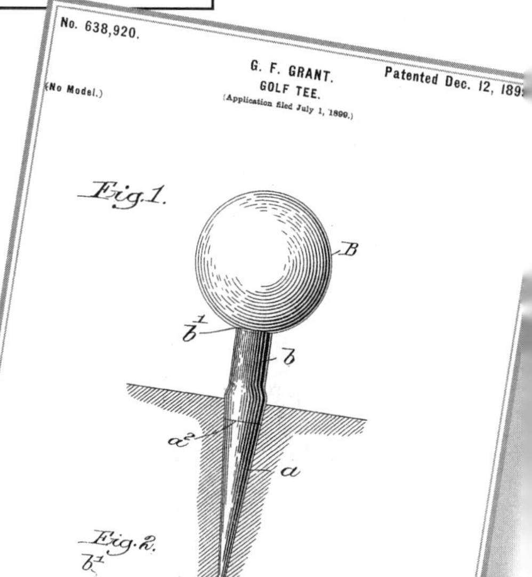

What is it with dentists and golf tees? The earliest wooden tee patented in America was invented by George Grant, the first African-American graduate of the Harvard Dental School. It was a rubber tee with a wooden top. He had a small number made for himself and his friends, but they never caught on with the greater golfing public.

1923

FISH OUT OF WATER

A flash of inspiration

In 1913, a young biologist set out on a dogsled journey across Labrador. It was his first trip to the frozen north, and he looked in wonder at everything he saw. One thing in particular caught his attention: "I saw natives catching fish in fifty-below weather, which froze stiff almost as soon as they were taken out of the water. Months later, when they thawed out, some of these fish were alive."

Later, he was joined by his wife and five-month-old child. Providing them with fresh food was a challenge. Observing the Inuits, he saw that food frozen quickly in the dead of winter stayed fresh and tasty for a long time.

Back at home, he began to experiment with freezing foods. He used a $7 fan, a bucket of ice, and a bucket of seawater. His wife complained about the live fish in the bathtub, but he persevered. In 1923, he invented a system of packing fresh food into cardboard boxes, and then flash-freezing them between metal plates.

Clarence Birdseye knew flash-freezing was a hot idea. He formed a company that began selling packages of frozen fish in a Springfield, Massachusetts, store in 1930. The public eventually warmed to the idea, and the frozen food industry was born.

Clarence Birdseye eventually held more than three hundred patents.

Birdseye spent five years traveling by dogsled in Labrador, and developed some adventurous eating habits, trying everything from blackbirds and starlings to lynx and lizards. "And I'll tell you," he said, "the front half of a skunk is excellent."

"I AM INTENSELY CURIOUS ABOUT THE THINGS WHICH I SEE AROUND ME."

—CLARENCE BIRDSEYE, EXPLAINING THE SECRET OF HIS SUCCESS

The key to frozen food is freezing it as quickly as possible. Freezing it slowly forms larger ice crystals that tear the food apart, making it a mushy mess when thawed.`

1923

THE HONEY BEE BOOGIE

Unlocking the secret language of bees

In November of 1923, a little-known agitator named Adolf Hitler tried to kidnap German leaders in a Munich beer hall, and take over the government. The event, which became known as the Beer Hall Putsch, was a prelude to Hitler's eventual rise to power.

Across town, a scientist named Karl von Frisch was doing his best to ignore the political goings-on in favor of something he found far more interesting.

He was watching bees dance.

Frisch was researching how bees find their way to flowers with nectar. He originally thought it had to do with scent. But years of painstaking observation led him to a startling conclusion: bees that find nectar can communicate its distance and direction through a dance. A slam dance, really: arriving back at the hive, the bee plunges into a crowd of other bees by the entrance and jostles them in narrow circles. The movements performed by the bee communicate highly accurate information to other bees on how to find the nectar.

Frisch's eye-opening discovery that bees have their own dance language inspired renewed interest in the field of ethology, the study of animals in the wild. And fifty years later it won him a share in the 1973 Nobel Prize in medicine/physiology.

Even after Frisch won the Nobel Prize, some scientists continued to question his hypothesis. They argued that it just wasn't possible for bees to communicate all that information through dancing, and that scent must play a role. But researchers in Britain have placed micro radio-transponders on honeybees and used scent-free "nectar" to confirm that the dancing is, in fact, the way the bees communicate.

A bee that has found nectar performs two basic dances to communicate distance and direction: the "circle dance" and the "waggle dance." Other bees try to touch the dancing bee, so that, in Frisch's words, "the dancer herself, in her madly wheeling movements, appears to carry behind her a perpetual comet's-tail of bees." Experiments show that the bees orient themselves using the sun, and that the flight data they communicate takes into account wind speed and direction.

1926

HAIR TODAY

Pinning down the latest fashion trend

When young women started cutting their hair short in the 1920s, it turned America upside down. Ministers proclaimed that women who bobbed their hair were flirting with evil. Mothers fretted they couldn't tell boys from girls—at least from behind. One teacher who bobbed her hair was told she would be fired unless she grew it back. Nurses were ordered not to bob. But nothing seemed to deter hordes of women from lopping off their locks. The Roaring Twenties and the age of the flapper soon followed.

The controversial fashion trend proved a boon for hairdressers. In 1921, there were 5,000 hairdressing shops in the United States. By 1926, there were 21,000. But it was a disaster for hairpin manufacturers. A woman with bobbed hair had little use for traditional hairpins, and the bottom fell out of the market.

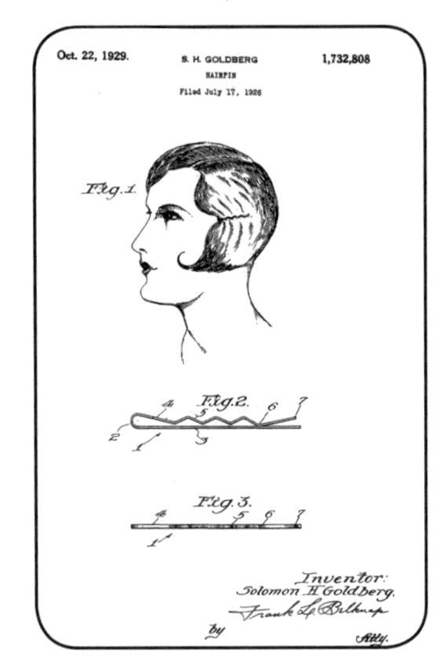

Into this moment of hair hysteria stepped Sol Goldberg, known as the "Hairpin King of Chicago." With several hairpin patents to his name, Goldberg was just the man to whip up something new: a pin with one straight leg and one crimped leg. The perfect pin to inconspicuously hold in place the sides of the fashionable new haircuts.

In his patent application, he said it was "designed particularly for use with short or bobbed hair." And it was bobbed hair that gave his soon-to-be-ubiquitous invention the name it has gone by ever since:

The bobby pin.

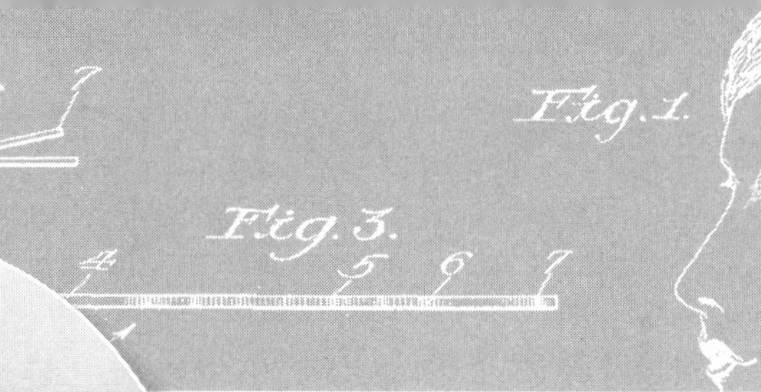

How did the bobbed hair craze of the 1920s get started? Some have traced it back to celebrated ballroom dancers Irene and Vernon Castle. Irene cut her hair short around 1915 for convenience's sake—it was easier to do all that dancing with short hair. One of the most popular variations of the short haircut became known as the "Castle Bob."

Numerous fictional characters have used bobby pins to pick locks or otherwise save the day. But truth almost always trumps fiction. In 1961, the pilot of an Australian airliner flying from Brisbane to Sydney used a bobby pin to short-circuit the electrical system of his malfunctioning landing gear. That enabled him to shake loose a jammed nose wheel and make a perfect landing. MacGyver would be proud.

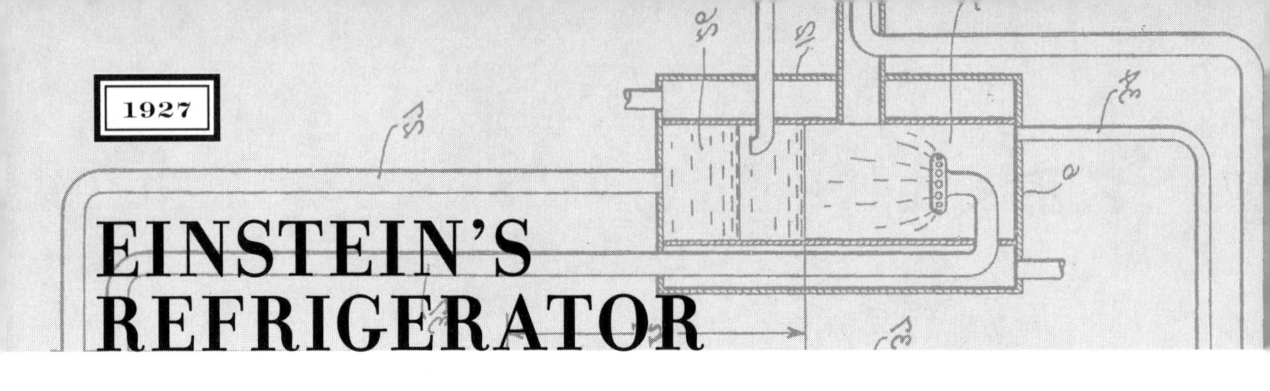

1927

EINSTEIN'S REFRIGERATOR

Genius on ice

Albert Einstein may be the most famous scientist in history. His very name is a synonym for genius. Less well-known is the fact that he invented a refrigerator.

Why did the world's greatest scientist want to work on something as prosaic as a kitchen appliance? In Einstein's day, refrigerators used ammonia. If the toxic gas leaked out, it could be fatal. Einstein read about the death of a family from a leaking refrigerator, and set out to make something safer. At least, that's how one version of the story goes. Another says he was just trying to invent a quieter fridge—imagine a refrigerator so loud even a genius can't hear himself think!

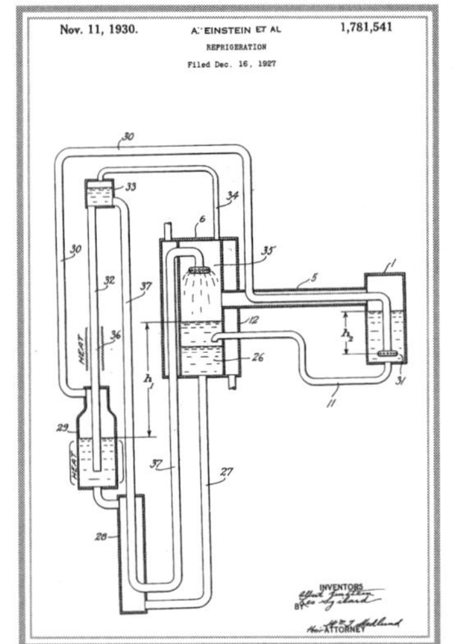

Einstein and a former student, Leo Szilard, invented a refrigerator with no moving parts, and thus less potential for leakage. It featured a revolutionary electromagnetic pump. They also invented a device for cooling a beverage, which was powered by the pressure of a water faucet. They obtained forty-five patents in six countries for their ideas.

Einstein and Szilard failed to corner the refrigerator market, but maybe they were just ahead of their time. As of 2008, scientists at Oxford University are using the principles of the Einstein-Szilard refrigerator to come up with a new kind of fridge that is more environmentally friendly.

Now that's cool.

Leo Szilard went on to become a renowned scientist on his own. He used some of the money he made from the sale of the refrigerator patent rights to fund his later work on atomic energy. The pump he and Einstein invented was later used in nuclear reactors.

The refrigerator wasn't Einstein's only foray into inventing. He also collaborated with German inventor Rudolf Goldschmidt to invent a hearing aid. Einstein actually wrote to Goldschmidt in verse to beg him to work on the project.

1928

THE MOLD THAT SAVED MILLIONS

How did a messy research lab lead to the development of a wonder drug?

There was nothing unusual about the fact that Scottish biologist Alexander Fleming failed to clean up his lab before going on holiday in the summer of 1928. Friends often teased Fleming for being disorderly. The truth is that he was very hesitant to throw out his old bacteria cultures until absolutely sure that there was nothing more to learn from them.

He came back from vacation to find some petri dishes had grown moldy. Sorting through them prior to throwing them out, he discovered that the mold in one dish had destroyed the bacteria culture he was growing there. The mold was a kind of fungus, penicillium, that grows on bread. Fleming wrote a scientific paper on his discovery, but never really followed up on its practical applications.

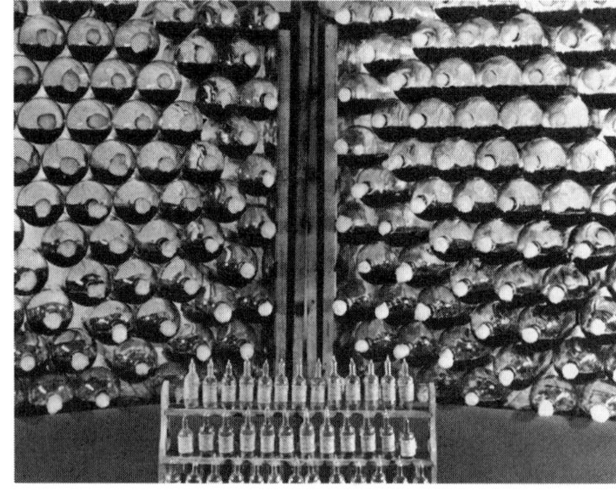

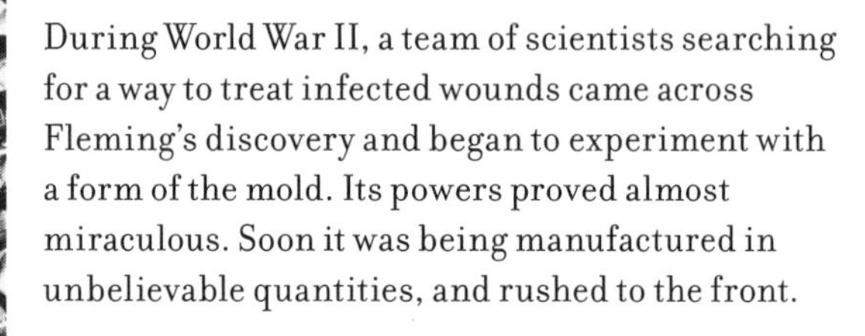

During World War II, a team of scientists searching for a way to treat infected wounds came across Fleming's discovery and began to experiment with a form of the mold. Its powers proved almost miraculous. Soon it was being manufactured in unbelievable quantities, and rushed to the front.

More than fifty years later, penicillin remains the world's most used antibiotic. Thanks to a scientist who didn't like to clean up.

“THAT’S FUNNY . . .”

—FLEMING, HIS VOICE TRAILING OFF, WHEN HE DISCOVERED THE MOLDY DISH

More than twenty chemical companies participated in a crash program to manufacture lifesaving penicillin during World War II. By war’s end, they were manufacturing 650 billion units per month.

Fleming shared in a Nobel Prize for his serendipitous discovery, about which he dryly commented: “One sometimes finds what one is not looking for.”

1929

REMEMBERING BILL LEAR

The restless inventor who put music on wheels

Why should we remember Bill Lear? To begin with, he was a high school dropout who nonetheless earned more than 150 patents.

If you like to listen to music in your car, you have two really good reasons to remember Bill. In 1929, he helped invent the world's first practical car radio: the Motorola. (It took its name from the words "motor" and "Victrola.") Thirty-five years later, he returned to the field of mobile music, patenting the eight-track tape player, which filled millions of cars with music in the 1960s and 1970s.

That's not enough? Lear was also a pioneer in radio navigation and autopilot technology for airplanes. In 1950, President Truman awarded him the Collier Trophy, one of aviation's highest prizes, for an autopilot capable of safely landing a jet-fighter on a fogged-in runway.

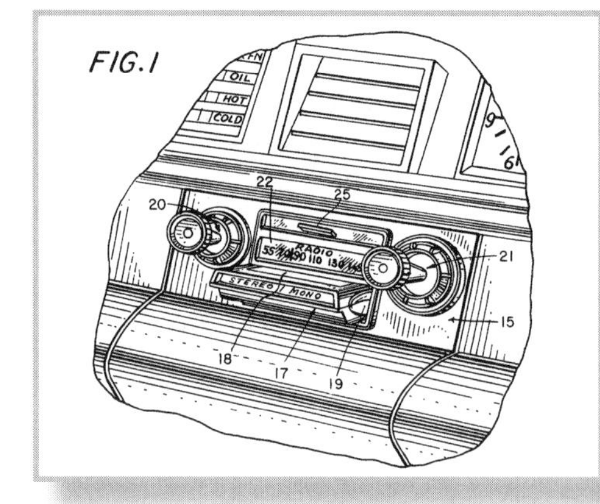

Those are a few of the reasons we should remember Bill Lear. The reason we *do* remember him is that he adapted a Swiss military jet into a high-performance luxury aircraft for business executives. The Learjet made his name a household word.

Lear never quit tinkering. A few days before his death at age seventy-five, his wife came to visit him at the hospital. "I don't have time for this nonsense," he told her. "I have work to do."

In the 1970s, Lear set out to solve the problem of air pollution by inventing a low-emission steam engine that would power the cars of the future. He built a steam-powered bus with the words "steam is beautiful" on the top. But the idea never got rolling.

Another reason to remember Bill Lear is his daughter Shanda. If that seems unremarkable, say her name out loud: Shanda Lear. She overcame that to build a career as an entertainer and motivational speaker.

Lear's eight-track was a re-engineering of a four-track tape system invented by Earl "Madman" Muntz, a Los Angeles entrepreneur who became famous selling cars and consumer electronics with outrageous TV commercials playing up his "Madman" personality. A typical line: "I wanna give 'em away, but Mrs. Muntz won't let me. She's crazy!" Muntz is credited with coining the shorthand "TV" for television , and he too showed a sense of whimsy in naming his daugher, christening her TeeVee.

1935

METER MAN

The feisty newspaperman who made drivers hate him

Carl Magee was a crusading Santa Fe newspaper editor who helped expose the famous Teapot Dome oil scandal. His blistering editorials embroiled him in libel suits and fistfights. When an angry judge attacked him on the street, Magee pulled his gun. He winged the judge but killed a bystander. Tried for manslaughter, he won an acquittal on the grounds of self-defense. So this rough-and-tumble character was no stranger to controversy by the time he invented one of the most unpopular contraptions of all time:

The parking meter.

In the 1930s, Magee moved to Oklahoma City. While working as the editor of the *Oklahoma News*, he headed up the local chamber of commerce's traffic committee. Parking congestion was hurting businesses, and nothing the police tried seemed to work. Magee's answer: a coin-operated timing device on the curb. He worked with a graduate student at the local engineering school to perfect a model, and persuaded the city to let him test it out.

On July 16, 1935, 174 meters went into operation. Angry drivers immediately staged a revolt, taking the issue to court. A judge issued a temporary restraining order, and it looked like time might be running out for Magee's meters. But the judge eventually ruled in Magee's favor. (Possibly he didn't want to get shot?)

Magee's meters soon spread to other cities. Within two years, more than twenty thousand had been installed. Meter maids and the boot followed. Thanks a lot, Carl.

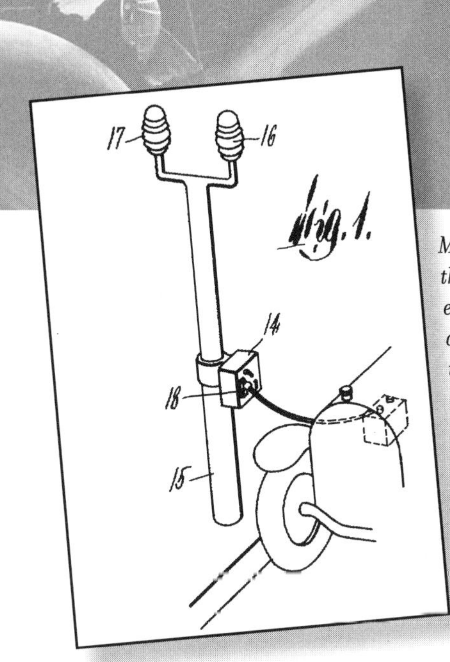

Magee's parking meter was not the first to earn a patent. One earlier idea called for a timing device that would be attached to a car. Another, patented by Roger W. Babson, founder of Babson College, envisioned drivers plugging their car battery into a curbside light fixture that would start flashing when the car had overstayed the time limit.

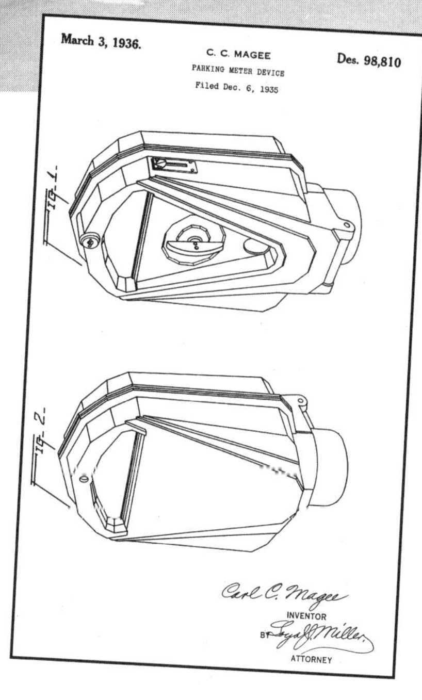

Magee called his invention the Dual-Park-O-Meter, because it served a dual purpose: controlling parking and bringing in money for the cities.

Drivers took heart when the Alabama Supreme Court outlawed Birmingham meters as "an unauthorized exercise of the taxing power" in 1937. But it was a temporary victory in a losing war.

DIABOLICAL RAYS

The invention that saved Britain

In 1924, a British inventor named H. Grindell Mathews claimed to have invented a death ray. He announced that his "diabolical rays" could knock down a fleet of airplanes, and that the Germans almost certainly possessed the same dangerous technology. But Mathews was never able to prove that his death ray could do more than generate headlines.

Nevertheless, there continued to be widespread fear that Germany was developing a powerful death ray. After Hitler came into power, this became a pressing concern for Great Britain.

Officials at the British Air Ministry turned to a scientist named Robert Watson-Watt, who was experimenting with radio waves to locate storms. In February of 1935, they asked him to report on the feasibility of a radio death ray. He responded with calculations demonstrating that it was an unworkable fantasy. But in the same memo, he suggested what he called, with typical British understatement, a "less unpromising" possibility: using radio waves to pinpoint the location of enemy airplanes.

An idea that became known as radar.

Watson-Watt and his team were forced to work at breakneck speed. Testing began just weeks after that initial memo. He was the first to admit that the radar system they developed was imperfect, but that was in accord with his oft-expressed philosophy. "Give them the third best to go on with," he said. "The second best comes too late, the best never comes."

The government promptly forgot about death rays and put Watson-Watt to work developing his idea. Then they launched a crash program to build a system of radar stations along the coast. When Nazi bombers launched an all-out assault in 1940, radar gave British fighter planes a critical advantage. The "Battle of Britain" was a turning point in World War II, won because of an idea generated by a mythical weapon.

Death rays such as Mathews's (seen below) were popular newspaper fodder in the 1930s. In 1934, a scientist who had fled Nazi Germany for France claimed to have invented a handheld death ray that could stun people from a mile away. Famed inventor Nicola Tesla told reporters in 1935 that he had invented a death ray that could shoot down ten thousand airplanes at a distance of 250 miles.

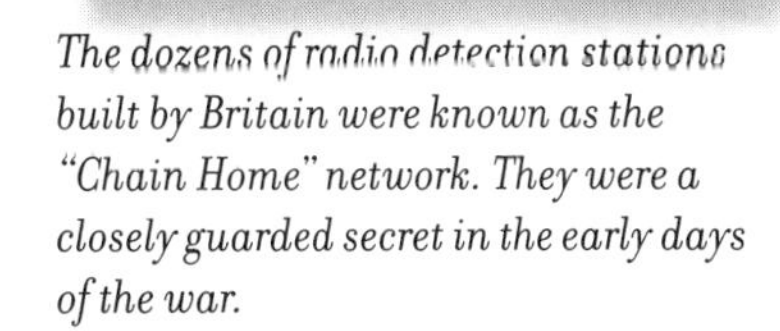

The dozens of radio detection stations built by Britain were known as the "Chain Home" network. They were a closely guarded secret in the early days of the war.

BENT OUT OF SHAPE

The power of being flexible

The inciting incident took place one summer afternoon at the Varsity Sweet Shop in San Francisco. Joseph B. Friedman was watching his two-year-old daughter try to drink a glass of soda at the counter. She bent the straw down so she could reach it, but in bending the straw, she cut off the flow.

Julia was struggling, and she wasn't happy.

Friedman was a real estate broker and part-time optician. But he was also a mechanically minded sort who came up with his first invention at fourteen and his first patent at age twenty-two. He had moved away from inventing to support his family, but in his daughter's frustration he spied an opportunity he couldn't pass up.

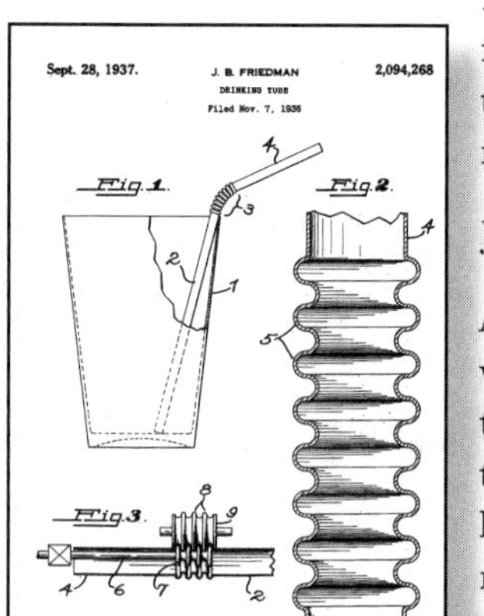

Friedman went home and inserted a screw into a straw. He wrapped dental floss around the grooves and then pulled it tight. This created corrugations that gave the straw a hinged elbow. Friedman filed a patent for his improved "drinking tube" in November of 1936.

Joe Friedman invented the flexible straw.

And his fortune was made, right? Not right away. Straw manufacturers weren't interested. At least one thought the new straw would be impossible to make. But Friedman wasn't about to give up. He made improvements to the straw, and decided to go into business himself. As simple as the straws look, it took more than a decade of work before he was ready to start making and selling them.

And *then* his fortune was made.

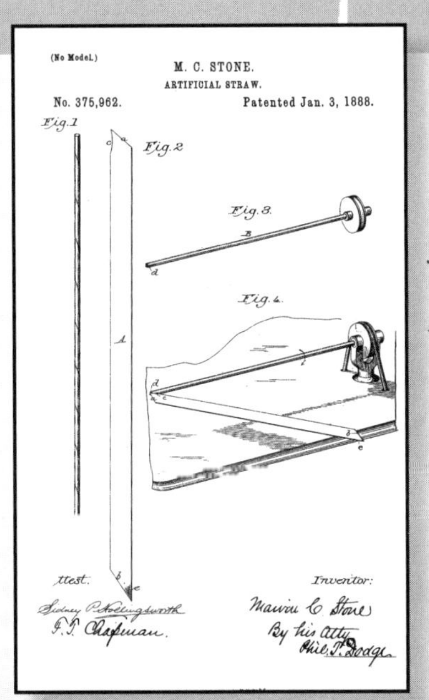

Marvin Stone supposedly invented the paper drinking straw while sipping a mint julep after work. He was using a straw made from a piece of rye grass, and finding it less than satisfactory. (Likely it was the grassy taste imparted by the straw that bothered him.) Stone, who had a business making cigarette holders, wound a strip of paper around a pencil, glued the ends together, and pulled out the pencil. He patented his paper straw in 1888.

Joe Friedman was constantly inventing. He ended up with nine patents and notebooks jam-packed with dozens of other ideas, including an early version of 3-D movies and a reflective license plate.

The Maryland Cup Corporation eventually bought Friedman's patents. Today the company makes 500 million flexible straws a year.

1937

ONE FOR THE ROAD

Drinking, driving, and the digital age

The modern computer was born in a roadside bar in Illinois one December night in 1937.

John Vincent Atanasoff was a physics professor at Iowa State University. He had been thinking about building a new kind of calculator that would be faster than the mechanical ones then in existence. To clear his mind one night, he went out for a drive. It turned out to be a long one—he wanted to get a drink, and since Iowa was dry, he had to drive more than two hundred miles to Illinois to imbibe.

He must have gotten a lot of thinking done on that drive.

Once at the bar, the ideas all seemed to come together. He hit on the idea of using the binary system—1s and 0s. He would create electronic circuits for calculating numbers—something that had never been done before. He came up with a way to constantly refresh the electrical charge in capacitors so he could use them to create memory banks. (All of these remain features of modern computers.) Feverishly, he scribbled everything down on a cocktail napkin.

Back in Ames, Iowa, he and a graduate student named Clifford Berry built the Atanasoff Berry Computer in the basement of the physics building. It was a primitive machine with two rotating drums of capacitors driven by a bicycle chain. It was so loud that everybody in the building could tell when it was turned on. But it was the first digital electronic computer—and a leap forward into a new age.

> "It was an evening of bourbon and 100 mph car rides when the concept came."
>
> —John Vincent Atanasoff

For many years, the ENIAC, unveiled at the University of Pennsylvania in 1946, was considered the world's first electronic computer. But it turns out that John Mauchly, one of the designers of the ENIAC, pirated some of Atanasoff's ideas after he examined Atanasoff's device in 1940. Atanasoff didn't have a patent—the lawyer for the university didn't understand the machine and left the undone paperwork in his drawer for years. But a 1973 court case eventually ruled that Atanasoff deserved the credit, because the ENIAC was derived from his work.

1938

A MAN CALLED VEEBLEFETZER

The man who made wireless work

In 1927, nine-year-old Al Gross sneaked into the radio cabin of a steamship on Lake Erie. "I heard the noise of the spark transmitter and I saw the radio operator and all of his radio gear. Boy did that impress me."

The inventing career it led to proved equally impressive.

At age twelve, Al became a ham operator. In 1938, at age twenty, he built the world's first handheld two-way radio: the walkie-talkie. His radios were used by the OSS during World War II to communicate with agents behind Nazi lines.

After the war, he became a pioneer in Citizens' Band (CB) Radio, building two-way radios for the public. Given the first CB license on a new frequency in 1958, he chose the "handle" Phineas Thaddeus Veeblefetzer.

Sometimes called the father of wireless communications, Gross created circuitry that led the way to cordless phones, cell phones, even the garage-door opener. In 1949, he invented the pager—but he had trouble marketing his latest invention. "The doctors hated it. They complained that it would interrupt their golf games."

Gross was often ahead of his time—CB and pagers didn't become popular until his patents ran out. His inventions never made him rich, except, as he once said a few years before his death in 2000, in experience. But he never seemed to care.

The thrill of inventing proved payment enough for the man called Veeblefetzer.

> **"If I still had the patents, Bill Gates would have to stand aside."**
>
> —AL GROSS

Gross built walkie-talkies used in the OSS "Joan-Eleanor" project. His radios weighed less than four pounds . . . astonishing in an age of vacuum-tube technology. They needed to be small because undercover agents inside Nazi Germany were secretly using the radios to communicate with Allied planes flying overhead.

In the 1940s, cartoonist Chester Gould visited Gross and found him tinkering with the idea of a portable radio so small you could wear it on your wrist like a watch. Gould asked if he could borrow the idea for his cartoon detective Dick Tracy. The two-way wrist radio became an iconic element of the strip.

THE MYSTERY OF THE VANISHING VAPOR

A slippery story of hide-and-seek

DuPont chemist Roy Plunkett couldn't figure it out. There was supposed to be gas in the canister. But when his assistant, Jack Reebock, opened the valve, nothing came out. Someone else might have just assumed the canister was leaky or defective, and thrown it out.

Good thing these guys took a different approach.

The two men were working on coming up with a nontoxic refrigerant that could replace Freon. The gas was a compound called tetrafluoroethylene (TFE) that they had mixed up and were keeping on dry ice.

They weighed the canister and found it weighed just what it should with the gas inside it. They stuck a wire in the valve to make sure it wasn't plugged up. Still mystified, they did the only thing they could think of.

They sawed the canister in half.

Inside, they found that the gas had morphed into a greasy white powder. Plunkett sniffed it, tasted it, and began testing it. He came to realize that he had accidentally created the most slippery substance on earth. Something that would one day be used in everything from the atom bomb to frying pans.

Teflon.

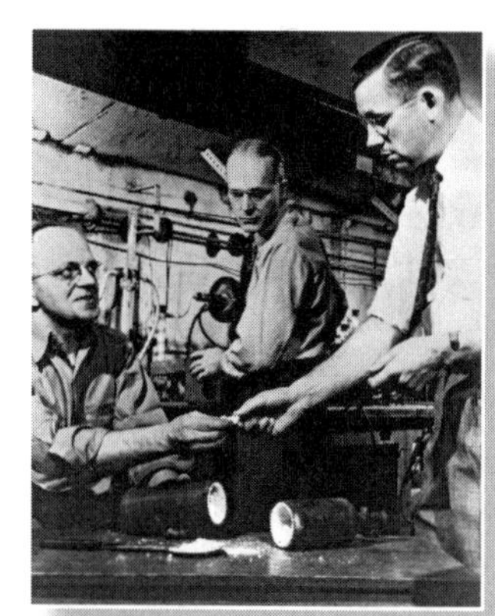

Plunkett and Reebock re-create the experiment that started it all.

In addition to being slippery, Teflon is also inert—no other substance will chemically react to it—and it is virtually impervious to heat.

A French engineer named Mark Gregoire had the idea of coating his fishing gear with Teflon to prevent tangles. His wife suggested instead that he find a way to put the stuff on frying pans. After he did so, they formed the Tefal corporation and began selling the pans in France in 1956. Nobody in the United States seemed interested in Teflon cookware until Macy's put a few Tefal pans on sale in December of 1960. In spite of a severe snowstorm, they sold out instantly, and the age of the nonstick pan was born.

Teflon can also be found in bridges, space suits, rain jackets (Gore-Tex uses Teflon), surgical thread, and the interior iron framework of the Statue of Liberty, which is coated with Teflon to keep it from corroding.

1938

THE MAGIC BOX

The bright idea that spawned a trillion copies

Chester Carlson desperately wanted to invent something. Instead, the Caltech grad was stuck working in the patent department of a law firm. He spent hours making lists of possible inventions, including a toothbrush with replaceable bristles and an improved cap for soda bottles.

All the time, the answer to his quest was right there in his office.

Patent applications required numerous exact copies of technical drawings, each of which had to be made by hand. It was especially difficult for Carlson, because of his arthritis. How wonderful, he thought, to have a magic box that would let you push a button and get a copy.

He set out to come up with such a device. Even though he was working full-time *and* going to law school at night, he turned his kitchen into a lab and used every spare hour for experimentation. His chemicals caught fire on the stove and sent foul-smelling smoke through the apartment, but he was undeterred.

In 1938, he and an assistant made their first successful copy. Carlson took his idea to IBM, GE, and others, but no one was interested. His assistant quit. His wife left him. Finally, the Haloid Company of Rochester, New York, took a gamble on his idea. They poured millions into making a workable machine.

Needing a catchy name for the process, Haloid took the Greek words *xeros*, for "dry," and *graphos*, for "writing," to make a new word: "xerography." And the company changed its name: to Xerox.

10.-22.-38
ASTORIA

Carlson and his assistant Otto Kornei made their first successful dry copy (above) on October 22, 1938. It was called a "dry" copy to distinguish it from photographs that needed to be developed in a bath of chemicals. Carlson's process involved projecting an image on a negatively charged plate, so that positively charged particles of powder stuck to the dark areas but not the light areas. They baked that powder onto a piece of wax paper to make the copy.

Chester Carlson became a multimillionaire. When his second wife asked him if there was anything else he wanted, he told her: "Yes, I'd like to die a poor man." He gave away 100 million dollars to charity before his death.

Xerox Model A went on the market in 1949. The "Ox Box," as it was soon nicknamed, took forty-five seconds to make a single copy. After another decade of tinkering, Xerox came out with the Model 914 in 1959 and took the country by storm.

1941

A REAL WONDER

The truth about William Marston

William Marston was a graduate student doing research in psychology at Harvard in 1915 when he developed what he called "The Marston Deception Test." It involved plotting a subject's blood pressure on a graph while asking them questions. Lies, he said, would show up as discontinuous data on the graph.

In other words, William Marston invented the lie detector.

Marston always said he invented a process, not a machine, but he earned international fame as the father of the polygraph. Although he believed it infallible when correctly administered, its accuracy has been a source of controversy and disagreement for much of the last century.

But the lie detector is not Marston's only claim to fame. Marston was a man of strong ideas who believed that society was doomed unless women eventually took over. So, he decided to create a strong hero-figure for young women. "Not even girls want to be girls so long as our feminine archetype lacks force, strength, and power," he wrote. "The obvious remedy is to create a feminine character with all the strength of Superman plus the allure of a good and beautiful woman."

And so it was that a quarter of a century after inventing the lie detector, Marston created a comic-book character who has fascinated us just as much as, if not more than, his first creation:

Wonder Woman.

Dr. Marston (administering a polygraph test at left) had an unconventional home life. He lived with his wife, Elizabeth, and his lover, Olive Byrne (both accomplished women), and had two children with each. The heavy silver bracelets that Olive wore became the inspiration for Wonder Woman's bracelets, which, in Marston's words, "protect her against bullets in the wicked world of men."

"WOMEN WILL WIN!"

—WILLIAM MARSTON

Wonder Woman first appeared in DC Comics in December of 1941. Marston wrote the comic strips until his death in 1948, including this 1943 story, in which Wonder Woman runs for president.

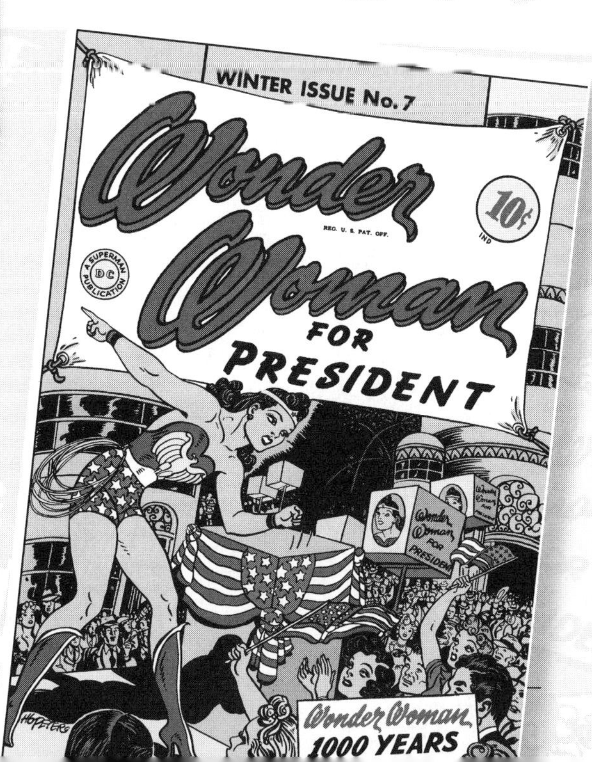

1942

THE UNSINKABLE SHIP

World War II Ice-Capades

During World War II, the Allied high command considered building an aircraft carrier made out of ice, which would have been the biggest ship in the world.

The project was the brainchild of Geoffrey Pyke, a brilliant but eccentric inventor working for the British military. Actually, the ship would have been built out of a sort of "super ice" called, in Pyke's honor, Pykrete. It consisted of a mixture of water and wood pulp that turned out to be incredibly strong and slow to melt. Plus you could saw and shape it just like wood.

The ship that Pyke conceived, the SS *Habbakuk*, would be a huge flat-topped iceberg, two thousand feet long and three hundred feet wide, with sides forty feet thick. It would be big enough to launch the biggest bomber, but virtually impervious to torpedoes and bombs. And the ship could carry refrigeration equipment allowing it to be self-healing.

British Prime Minister Winston Churchill became a big backer of the idea, and a prototype was successfully built in the Canadian Rockies. But critics, including U.S. Navy Commander Admiral Ernest King, argued that the project would absorb a vast quantity of war materials that could be better used elsewhere.

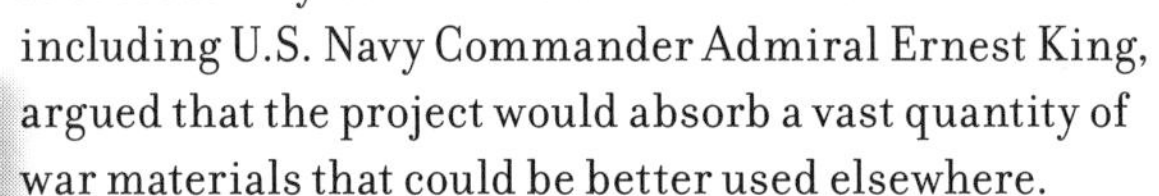

And so the idea of an iceberg warship melted away.

The prototype, which resembled nothing more than a houseboat, was built on a lake in the Canadian Rockies. It was sixty feet long and thirty feet wide.

An unusual demonstration of Pykrete's properties took place at a meeting of the Allied Joint Chiefs of Staff in 1943. Lord Louis Mountbatten, Chief of Combined Operations for the British military (far left), whipped out a pistol and fired a bullet into a piece of ice, which shattered. Then he fired at a piece of Pykrete, which was virtually undamaged. People outside the conference room were unnerved by the sound of shots going off. "My God," shouted one British officer, "the Americans are shooting the British." On another occasion, Mountbatten supposedly deposited a piece of Pykrete in Churchill's bathtub (while the prime minister was still in it) to demonstrate how long it took to melt.

> **"I ATTACH THE GREATEST IMPORTANCE TO THE PROMPT EXAMINATION OF THE IDEA."**
>
> —BRITISH PRIME MINISTER WINSTON CHURCHILL IN 1942 MEMO ON PYKRETE

The ship would have been twenty times bigger than the Queen Mary, *then the largest ship afloat. It was named after the Biblical prophet Habakkuk, who said: "I will work a work in your days, which ye will not believe." But the British Admiralty consistently misspelled the prophet's name, adding a* b *and taking away a* k.

1943

MIRACLE POISON

From chemical warfare to chemotherapy

During World War I, chemical weapons killed more than thirty thousand soldiers and injured three-quarters of a million. These poison gases were the first weapons of mass destruction. One of the worst was a deadly poison called mustard gas. It caused hideous burning and blistering, choking and blinding.

But this gruesome killer gave birth to an amazing lifesaver.

In the 1930s, the United States developed a new kind of mustard gas called HN2. Although the use of such gases had been banned by treaty, the United States felt the need to develop the weapons "just in case." The government contracted with Yale University to have researchers there look into the effects of the gas, and possible antidotes.

One of the things that Louis Goodman and Alfred Gilman discovered was that HN2 was an incredibly powerful agent for killing white blood cells, the fastest-multiplying cells in the body. It led them to wonder if it would also kill another kind of fast-growing cells.

Cancer cells.

Working first with mice, and then with humans, Gilman and Goodman showed that HN2 could retard the growth of certain tumors. It became the world's first effective cancer drug, and is still used today against leukemia and Hodgkin's disease. From a chemical weapons program came chemotherapy, and a powerful new weapon in the war against cancer.

A World War II German bombing attack exploded an American freighter carrying HN2 in the Italian port city of Bari in December of 1943. Six hundred and seventeen American military personnel were treated for mustard gas injuries, eighty-three of whom died, not to mention unknown numbers of Italians. Tragic as it was, the bombing gave researchers volumes of critical information about the effects of HN2 on the human body.

1944

A RHINO IN NORMANDY

Yankee ingenuity saves the day

Curtis Culin was a sales rep for a liquor distillery in New Jersey. He has only one invention to his credit. He never thought to patent it, and never made a dime off of it. But it is credited with helping the Allies overcome the Germans in the weeks following D-Day.

Once off the Normandy beaches, soldiers found themselves fighting in the farm fields. These were walled in by centuries-old hedgerows, dirt mounds three to five feet high and covered with thick shrubbery. They posed an unanticipated problem. A tank trying to drive over the hedgerow exposed its vulnerable underside. Ramming through took three or four attempts, eliminating the element of surprise. The Germans took advantage of this to slow the Allied offensive to a crawl.

That's when Sergeant Culin of the 102nd Reconnaissance Squadron had an idea for an attachment welded onto the front of a tank, which would channel the tank's power *under* the hedgerow. The squadron's maintenance officer worked with him to build a prototype out of a captured German antitank obstacle. The "Rhino" had four prongs with sharpened edges. A tank equipped with one could burst through a hedgerow in a single attempt.

The U.S. Army flew arc-welding crews over from England and set up a round-the-clock assembly line to outfit the tanks. Within a few days, more than five hundred were equipped. Allied Commander Dwight Eisenhower called it a "godsend." Sergeant Culin's invention helped the Allies fight their way out of the confines of Normandy, and come one step closer to defeating Hitler.

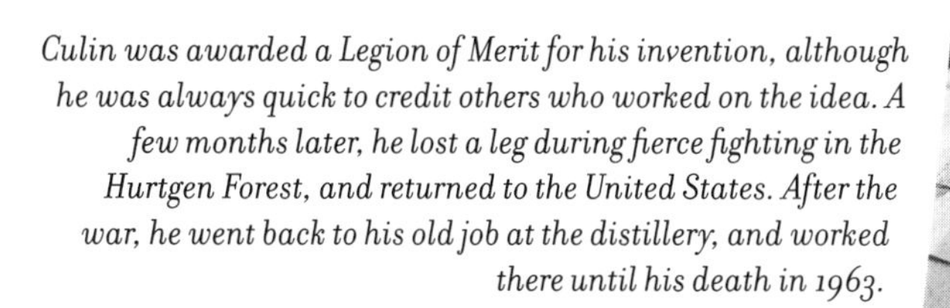

Culin was awarded a Legion of Merit for his invention, although he was always quick to credit others who worked on the idea. A few months later, he lost a leg during fierce fighting in the Hurtgen Forest, and returned to the United States. After the war, he went back to his old job at the distillery, and worked there until his death in 1963.

"There was the little sergeant. His name was Culin, and he had an idea."

—General Dwight D. Eisenhower

1945

COOKING WITH RADAR

A kitchen essential that was essentially an accident

In the summer of 1945, engineer Percy Spencer was conducting tests on a magnetron. That's the powerful tube at the heart of every radar set. When he reached into his pocket for a chocolate bar, he found instead a gooey mess. He wondered if the magnetron could be responsible.

Spencer was an engineering genius who had already helped win World War II by devising an improved magnetron tube that was easy to mass-produce, making possible the manufacture of tens of thousands of radar sets. Now he was ready to make his contribution to postwar America. Curious to see just what was going on, he put a bag of corn kernels in front of the magnetron. Voilà: the first-ever batch of microwave popcorn.

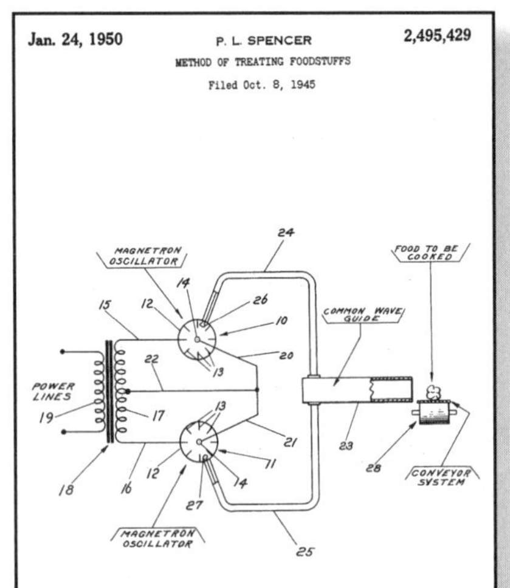

Spencer patented the new method of cooking. His employer, the Raytheon Corporation, transformed his discovery into the "Radar Range." The earliest model weighed 750 pounds, and had a price tag of $3,000. Sales were limited. It took more than twenty years for this "Big Daddy" of all microwaves to spawn a new generation that today graces kitchens everywhere.

> **“WHO’S GOING TO PAY $500 FOR A HOT-DOG WARMER?”**
>
> —EXPERTS CLAIMING THE MICROWAVE WOULD NEVER SELL

Raytheon’s first Radar Range was so big and expensive it only made sense in places like hotel kitchens and railroad dining cars. Today nine out of ten American homes have a microwave . . . albeit one a good deal smaller than the original.

Percy Spencer never had more than a third-grade education, but this self-taught engineering legend wound up with more than 150 patents.

1947

THE SWEET SMELL OF SUCCESS

An idea that was the cat's meow

Kaye Draper of Cassopolis, Michigan, inspired the development of a billion-dollar industry by asking her neighbor for a favor in January 1947. She wondered if he had any sand or sawdust she could use in her cat box.

That neighbor, Edward Lowe, operated a small business that sold absorbent materials to factories to clean up industrial spills. He didn't have any sand or sawdust, but he had a bag of absorbent clay called fuller's earth in the trunk of his car. He suggested that she try some of that.

Kaye came back a few days later asking for more. In addition to its absorbent properties, the fuller's earth absorbed odor. So her litter box didn't smell.

A lightbulb went on in Lowe's head.

He filled some five-pound bags with fuller's earth and wrote the words KITTY LITTER on them. Then he took them to a pet store, suggesting they sell them for 65 cents apiece. The owner hooted—no one would pay that much for a bag of clay when sand was a penny a pound.

So Lowe suggested he try giving them away.

Soon customers were coming back to the store asking for more. Lowe's brand-name litter products, Kitty Litter, followed by TidyCat, helped him build up a business that he sold in 1990 for about $200 million dollars.

Thanks in part to Lowe's innovation, cats passed dogs as the most popular pets in America. According to the American Pet Products Association, as of 2006 there were 88 million cats and 74 million dogs in the United States.

"I LOVE CATS."

—EDWARD LOWE

Lowe maintained a research center known as the "All American Cattery," where 120 felines who tested new products lived the high life. The facility boasted a fully staffed cat-care clinic, and twenty-four-hour television monitoring of resident cats.

Edward Lowe's spectacular success turned him into a conspicuous consumer who acquired, among other things, more than eleven thousand acres of land and a private railroad.

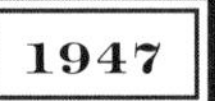

THE ADMIRAL AND THE INSECT

When their lives came together, a new term was born

On September 9, 1947, a luck-deprived moth flew inside a giant navy computer at Harvard. Alighting for a brief rest, the moth found its life irrevocably cut short when the relay upon which it had landed closed, crushing it between the contacts.

Undeniably unfortunate as this was for the moth, it didn't help the Mark II computer much, either. It crashed. Researchers crawled inside the monster machine (fifty feet long and eight feet high) to search for the problem. One of them retrieved the moth from the relay and, adding insult to (fatal) injury, taped it inside a logbook. It was the world's first computer bug.

Well, sort of.

Actually, the word *bug* had been in use for years to describe a minor technical malfunction. But the discovery of an actual bug in the machine tickled one of the computer's programmers, a hotshot navy mathematician named Grace Hopper. She adopted the term with a passion, drawing cartoons of computer bugs, frequently retelling the moth story, and writing a glossary for programmers that coined the term "debug." Thanks in large part to Hopper, the word became common currency among computer techs.

Hopper retired a navy admiral after a remarkable career in computing. But her name will forever be linked with the moth in the Mark II, and the computer bugs that plague us all.

A math professor at Vassar, Hopper joined the navy in World War II. After the war, she joined the Naval Reserve, and became part of the team developing the UNIVAC, the first commercial computer produced in the United States. She was such a force in computing that at age sixty-one the navy brought her back to active duty, and made an exception to its rigid retirement rules in order to keep her. She didn't retire until she was seventy-nine.

The use of the word "bug" goes back to Thomas Edison. A certain kind of interference in telegraph lines was referred to as a "bug," and when Edison came up with something to fix the problem, it was informally called a "bug trap."

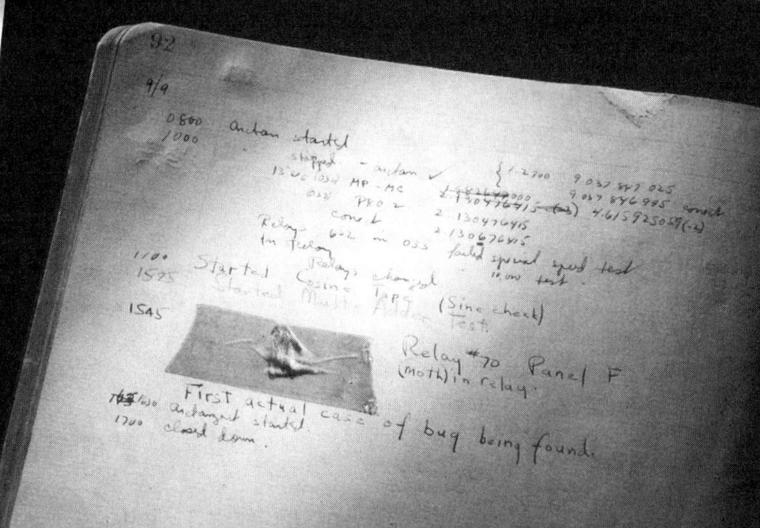

1948

HEAD CASE

It's all downhill from here

Howard Head desperately wanted to be a writer. When that didn't work out, he went to work as an engineer, designing aircraft parts during World War II.

After the war, he took up skiing as a hobby. It turned out he wasn't any better a skier than a writer. Like many others who have done poorly on the slopes, he put the blame on his equipment. He thought there must be a way to make a better ski than the heavy wooden ones then in use.

Head set out to make a lighter ski out of aluminum. But his early pairs kept shattering on the ski slopes. "Each time one of them broke, something inside me snapped with it," said Head.

But Howard Head was not about to give up on this dream. He quit his job and went to work full-time on the skis. His first pair gave way to his tenth and then his thirtieth. Finally, he took pair number forty to Mount Washington, and watched as an instructor rocketed down the slope. The excited instructor skied to a stop right in front of him.

"Howard, baby, I think you've got it."

Howard Head's aluminum skis revolutionized the sport, bringing skiing to the masses—and making a writer-turned-engineer a very rich man.

Head took a Polaroid of his first pair of skis. They were made of aluminum on the outside, with plywood on the inside. He pressure-cooked them in a tank filled with dirty motor oil to get the wood and aluminum to adhere.

Head financed his work on the skis with his life savings—six thousand dollars in poker winnings. When the skis kept breaking and money ran short, he despaired of succeeding. "The last month has felt like one long grind of grief." But his despair only drove him to work harder.

After his retirement, Head took up tennis—and once again decided that inferior equipment was getting in the way of his game. So he created the oversized tennis racquet that became today's standard.

1948

WALK THE DOG

Man's best friend inspires one of mankind's most ubiquitous inventions

Walking the dog isn't usually a life-changing activity. But it turned out to be exactly that for Swiss engineer George De Mestral. Returning from his canine constitutional one day, De Mestral noticed that both he and the dog had numerous cocklebururs sticking to them—souvenirs from hiking through a patch of heavy brush.

Nothing particularly unusual in that. But instead of just picking the cockleburs off and forgetting about them, De Mestral got to wondering what it was that made them stick so fiercely. So he stuck one under a microscope and looked at it. The burr had thousands of tiny hooks that grabbed onto anything nearby. It was Nature's way of ensuring that the seeds contained in the burr would get spread around.

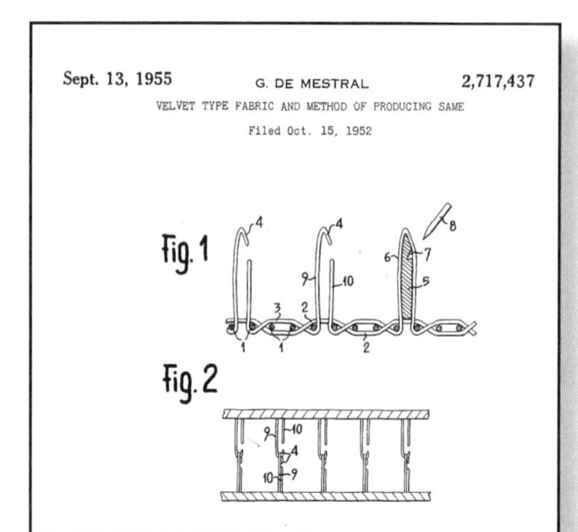

De Mestral wondered: Would it be possible to make a fastener with similar properties?

As the world would soon discover, the answer was a resounding yes. The invention he finally perfected in 1952: Velcro. Used today in everything from shoes to surgery to space travel.

George De Mestral's attachment to his dog helped change the way the world attaches things together.

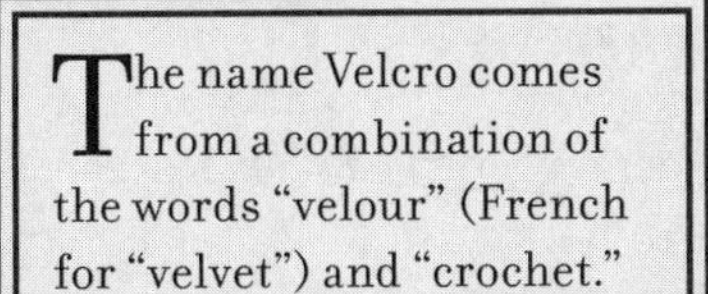

The name Velcro comes from a combination of the words "velour" (French for "velvet") and "crochet."

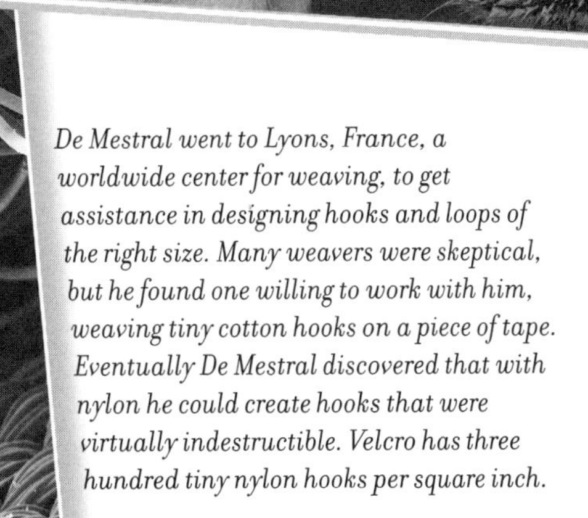

De Mestral went to Lyons, France, a worldwide center for weaving, to get assistance in designing hooks and loops of the right size. Many weavers were skeptical, but he found one willing to work with him, weaving tiny cotton hooks on a piece of tape. Eventually De Mestral discovered that with nylon he could create hooks that were virtually indestructible. Velcro has three hundred tiny nylon hooks per square inch.

BIG HANG-UP

1949

BEAUTY AND THE BOATER

Paving the way for the disposable diaper

Marion Donovan left her job as assistant beauty editor at *Vogue* magazine to raise a family. When her second daughter was born, she started wondering where the beauty had gone. Babies wetting their cloth diapers, the diapers soaking through sheets, everything constantly needing washing. She tried putting plastic baby pants around the diapers, but the elastic chafed the baby's legs and arms, and the pants caused rashes.

Donovan was raised in a family of inventors. One day she cut off a corner of the shower curtain and went up to her sewing machine in the attic to see if she couldn't fashion a waterproof diaper cover that would hold in the dampness without keeping the air out. Three years and not a few shower curtains later, she had a finished product. She called it the "Boater," because, she said, it helped keep babies afloat. And it had had an added benefit: it snapped together, eliminating the need for safety pins.

When they went on sale at Saks Fifth Avenue, they caused a sensation. "It is not often that a new innovation in the infant's wear field goes over with the immediate success of your Boaters," wrote Adam Gimble, President of Saks.

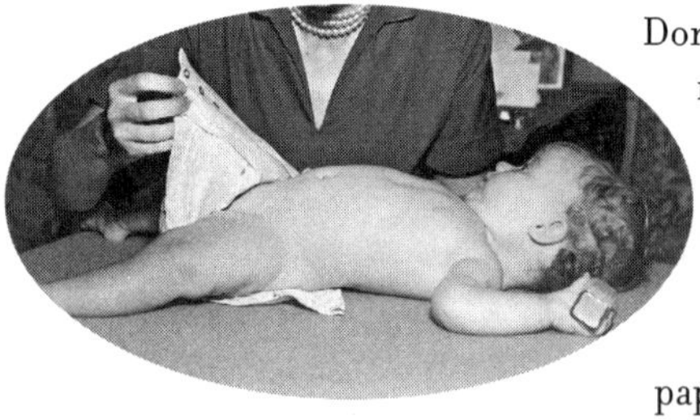

Donovan sold her patent for a million dollars, and moved on to her next big idea: a disposable diaper made of paper. She came up with a prototype, and pitched it to paper company executives.

But the execs—men all—thought the idea laughable. They couldn't see why in the world anyone would want them. And so it was another ten years before Procter and Gamble started papering baby bottoms with the Pampers.

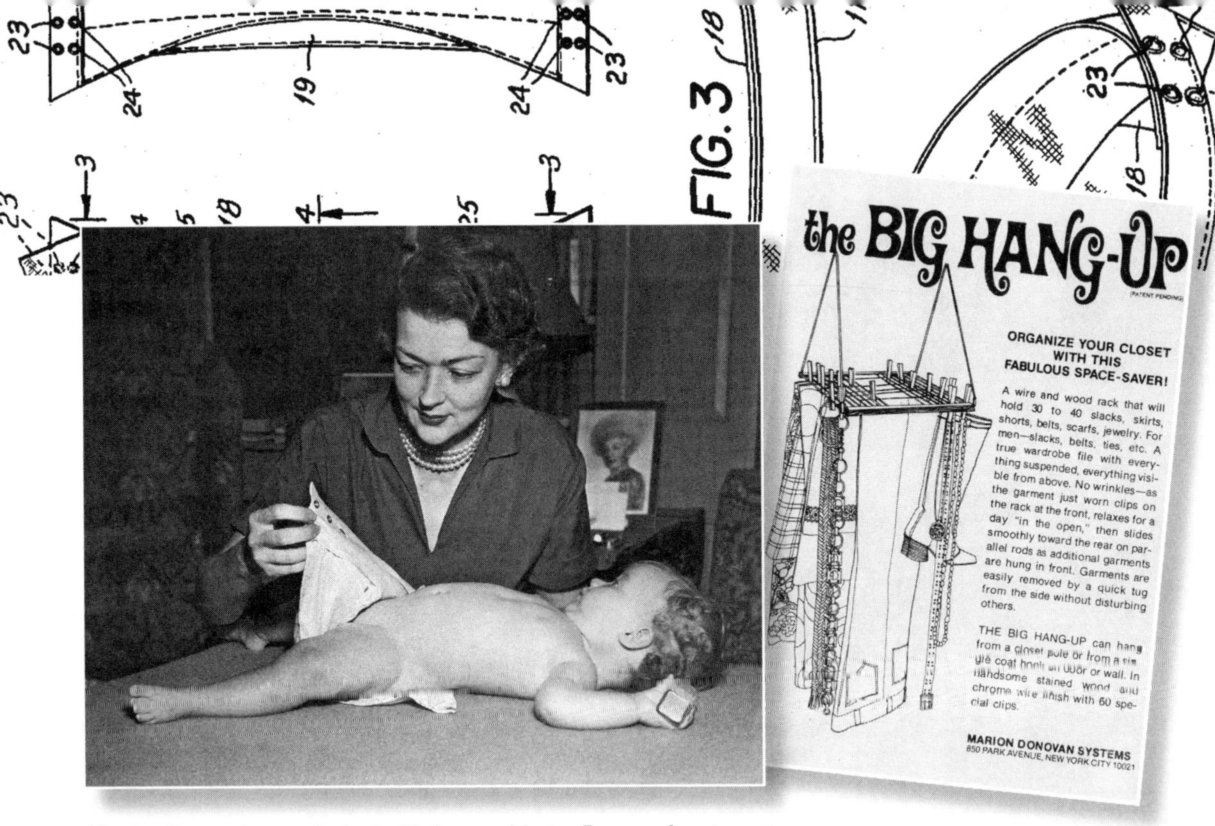

After her first product revolutionized baby care, Marion Donovan kept inventing, eventually obtaining more than a dozen patents for useful household thingamajigs. Among her other inventions are an elastic zipper called the "Zippity-Do," a closet organizer called "The Big Hang-Up," and precut pieces of dental floss called "DentaLoop."

Donovan was born into inventing. Her father, John O'Brien, and his twin brother, Miles, invented the South Bend Lathe, one of the most popular machine tools of the twentieth century. (Their company still exists.) After her mother died, her father often brought her to work, where she was entranced by the machinery.

1949

THE COOLEST THING ON ICE

Any guesses about what Frank Zamboni invented?

It is a precision machine with a top speed of nine miles per hour that fascinates sports fans of all ages. Seen in ice rinks around the world, it was born in the California desert.

In 1927, Frank Zamboni and his brother built a plant in Southern California to make ice for ice-boxes. Bad timing: the advent of the refrigerator soon threatened to put them out of business. So they decided to use their refrigeration machinery for a giant ice rink. In 1940, they opened Iceland. Iceland, as it was known, attracted hordes of skaters. But there was a problem.

It took a crew of five men armed with scrapers, shovels, squeegees, and a hose nearly an hour and a half to clean the ice. That brought the action to a screeching halt, and it was killing their business. The mechanically minded Zamboni decided to come up with a machine that could do the job quicker.

His first experiments ended in failure, but after the end of World War II he decided to try again. He used the engine from a Jeep, the axles of a Dodge truck, and some parts from a Douglas bomber. In 1949, his Model A ice-resurfacing machine was done.

And a new ice age was born.

Frank Zamboni had only a ninth-grade education, but he was a mechanical wizard with a strong stubborn streak: "I never would have finished it if people hadn't told me I couldn't do it," he said.

Skating sensation Sonja Henie helped the Zamboni ice-resurfacer really take off. When her Hollywood Ice Revue came to Iceland, she saw the Model A and decided she had to have one. Her troupe toured the country with their own Zamboni machine in tow, and soon every ice rink wanted one of their own.

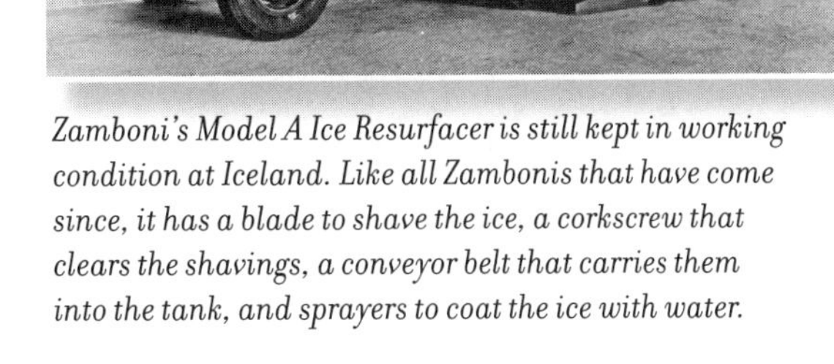

Zamboni's Model A Ice Resurfacer is still kept in working condition at Iceland. Like all Zambonis that have come since, it has a blade to shave the ice, a corkscrew that clears the shavings, a conveyor belt that carries them into the tank, and sprayers to coat the ice with water.

1955

EINSTEIN'S BRAIN

The strange odyssey of a genius's gray matter

In April of 1955, Dr. Thomas Harvey took the most famous brain in the history of the world and put it in a glass jar. Then he decided to keep it.

Harvey was filling in for the regular pathologist when he performed the autopsy on Albert Einstein at the University Medical Center in Princeton, New Jersey. He took the brain without permission, and was later dismissed from the hospital for doing so.

But somehow he managed to keep the brain. "I didn't know anyone else wanted to take it," he later said.

Harvey cut the brain into 240 fine sections and embedded it in cellodin, a substance used to preserve tissue samples, to allow for microscopic examination. But he never did any research on it himself. Over the years, he took it with him from New Jersey to Kansas and to California and back. He gave parts of it away to researchers, but kept most of it himself. For many years, he kept it in a jar behind a beer cooler in his office in Lawrence, Kansas.

More than forty years after taking the brain, the enigmatic Dr. Harvey decided to give it back. "Eventually, you get tired of the responsibility of having it," said Harvey, who gave the brain to Princeton pathologist Elliot Krauss in 1998.

Harvey (now deceased) has been attacked as a grave-robber and a crank. But thanks to him, Einstein's brain is still with us—and perhaps it will help future researchers unlock the secret of his genius.

Pathologist Thomas Harvey, talking to reporters in 1955 after performing the autopsy. More than forty years later, in 1996, Dr. Harvey asked two brain researchers in Canada if they wanted to study it. When they said yes, he drove across the country (he was in his eighties) with the brain in his trunk. The researchers calculated that Einstein's brain was smaller than normal, at 1.23 kilos, but that the inferior parietal lobe (the part that processes mathematical thought) was bigger than in most brains.

"In the future we might have the technology to . . . see what it is in his genes that made him so much smarter than the average man. We just plain don't have the knowledge now."

—Pathologist Elliot Krauss, the man who now has possession of Einstein's brain

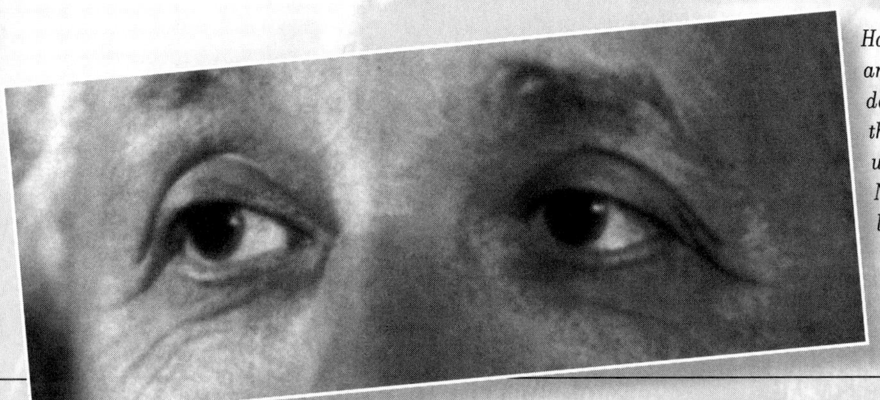

Harvey took out Einstein's eyeballs and gave them to the scientist's eye doctor, Henry Abrams. As of 2008, they are reported to be in an undisclosed safe deposit box in the New York City area. "When you look into his eyes, you're looking into the beauties and mysteries of the world," Adams is quoted as saying.

1956

SURFING SAFARI

Birth of the couch potato

Many people dislike TV commercials. Eugene McDonald wanted to obliterate them. The founder of the Zenith Corporation, McDonald thought it would be a better business model to have customers pay a subscription fee for commercial-free television. So he launched a war on ads, and set his engineers to work on a remote control that would help tune them out.

Their first effort, in 1950, was a remote connected by a cable called the "Lazy Bones." Customers didn't like the cable, so Zenith started working on a wireless remote. They had high hopes for the "Flashmatic," invented by Eugene Polley in 1955. It shot a light beam at a photoelectric cell on the TV. But there were problems. Televisions had a way of changing channels every time someone turned on a lamp.

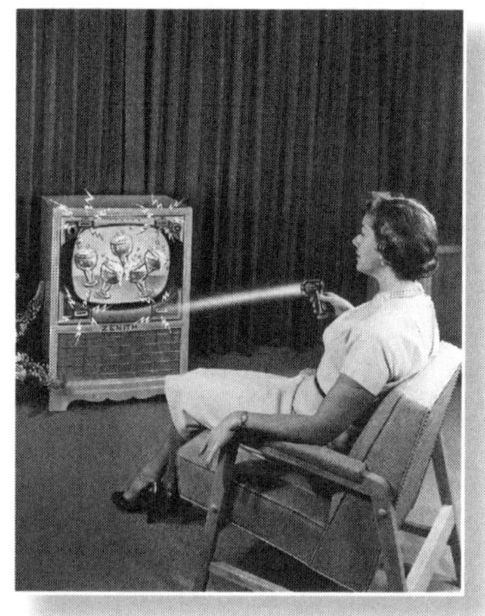

Back to the drawing board once more. And this time there was a demand from the Zenith sales force for a remote that didn't require batteries. Physicist Robert Adler came up with the answer: a channel-changer that used small, spring-loaded hammers to strike aluminum rods, giving off ultrasonic tones that could be deciphered by a receiver inside the TV. The remote made a clicking sound when you pushed the buttons, so it became known as a "clicker."

When McDonald saw it, he was ecstatic. "We gotta have it! We gotta have it!" he shouted. The new remote, called the Space Command, was introduced in 1956. Remotes failed to obliterate commercials, but instead gave birth to a nation of channel-surfers.

A prolific inventor, Adler was granted more than 180 patents. The first came in 1943, and the last was granted sixty-four years later, in 2007, just after he died. But according to his wife, the inventor of the remote wasn't much of a TV watcher. "He was more into books."

The remote itself was quite simple, with four buttons: channel up, channel down, mute, and power. But it required a receiver with six vacuum tubes inside the TV set to decipher the signals. It remained the industry standard for a quarter of a century..

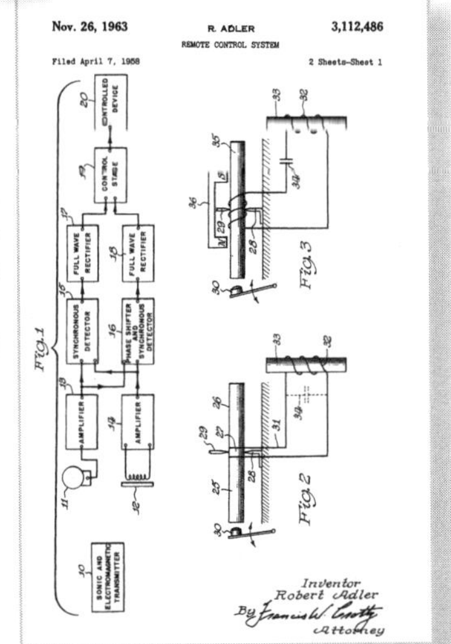

> **"People accuse me of creating the couch potato and I say, 'You're right!'"**
>
> —ROBERT ADLER, REMOTE CONTROL INVENTOR

1956

OOPS!

Mistakes happen . . . and sometimes it all turns out for the best

Bette Graham Nesmith wasn't a great typist. Which was unfortunate, because her job as an executive secretary at the Texas Bank and Trust required her to do a lot of typing. This in a day before word processors, when a single typing mistake could require starting the entire document over.

While earning some overtime decorating bank windows at holiday time, Graham noted how the artist she was assisting corrected mistakes not by erasing but by painting over. She put some white paint in a small bottle and used a tiny paintbrush to correct her mistakes.

She tried to keep her typing trick a secret, but coworkers eventually caught on. People began asking her to make bottles for them. Working nights and weekends, she turned her kitchen into a lab and her garage into a bottling plant. She recruited a chemistry teacher to help perfect the formula. She began selling hundreds of bottles a month—then thousands.

Graham called her product Mistake Out but eventually changed the name to Liquid Paper. Over the course of twenty years, she turned her one-woman venture into a major company that sold more than 65,000 bottles of Liquid Paper every day. She was eventually bought out by Gillette for nearly 50 million dollars, plus a royalty on every bottle sold.

Proving it is possible to earn from your mistakes.

Graham's teenage son Michael and his friends helped her fill the bottles in the garage. Michael Nesmith went on to fame as a member of one of the great TV/music sensations of the 1960s: The Monkees.

Bette Graham Nesmith was fired from her day job when she accidentally typed "The Liquid Paper Company" at the bottom of a letter instead of the name of the bank. She ran her business out of her house (and later a backyard shed) for more than a dozen years before finally investing in a factory and an office building.

1958

WHO INVENTED THE LASER?

You decide

One man won the Nobel Prize for it. Another holds several key patents—even though he didn't tell anybody about his research until years later. And a third actually built the first laser.

So, who invented the laser?

Columbia University professor Charles Townes was unable to sleep. He went for a walk, ending up on a park bench. Townes was searching for a way to produce extremely high frequency radio waves. Suddenly it came to him: stimulate atoms and molecules in a way that these tiny particles would emit the waves he was looking for. That insight eventually led to a Nobel Prize.

In 1958, Townes and another researcher laid out the theory behind the laser—and challenged someone to actually build it. Well-funded teams at places such as Bell Labs and IBM went to work, but the man who did it first was a maverick engineer at Hughes Aircraft named Ted Maiman. He described the laser he built in 1960 as "ridiculously simple."

Meantime, Chester Gould, a graduate student at Columbia University, had filled his notebooks with ideas for a laser without telling a soul. (He did get the notebooks notarized!) He dropped a patent application because he thought he needed a working model. But after twenty years of "laser wars" in court, he won patent rights that earned him millions of dollars.

So, who invented the laser? You decide.

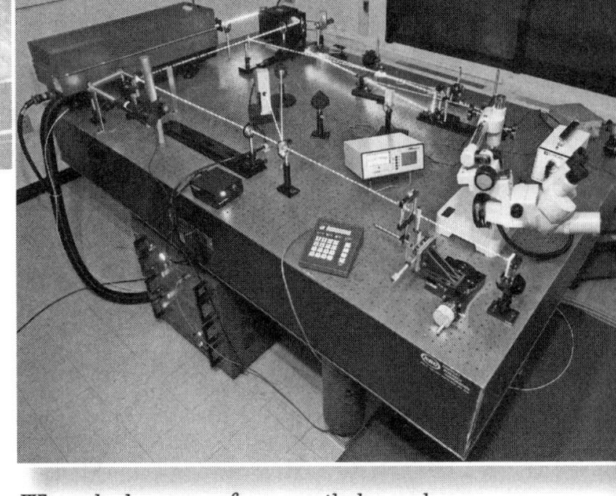

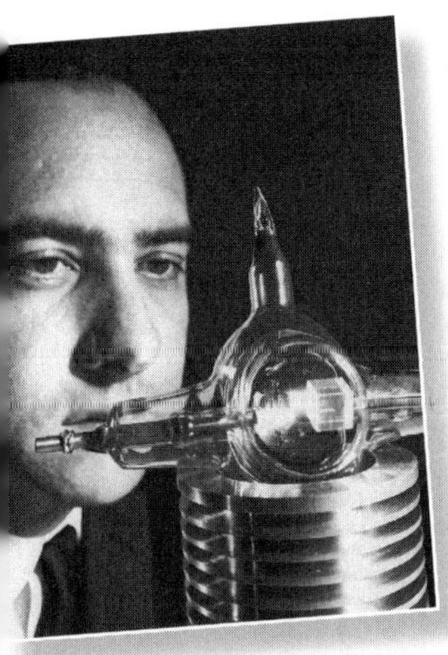

Maiman combined an off-the-shelf flash lamp and a ruby crystal. He never got the credit he thought he deserved, and he attributed it to what he called the "completion effect." After something is built, he said, it is easy to see how it was done—whereas before it is built, you not only don't know how to make it, but you don't even know for sure if it can be made. Up until his death in 2007, he refused to give his laser to the Smithsonian, instead keeping it in a shoebox under his bed.

When the laser was first unveiled, people wondered if its primary use would be as some sort of death ray. Today lasers are used for almost every purpose imaginable, from fiber-optic communications to computer disc drives, from eye surgery to architectural measurement. And, yes, they have also been used to shoot satellites out of the sky.

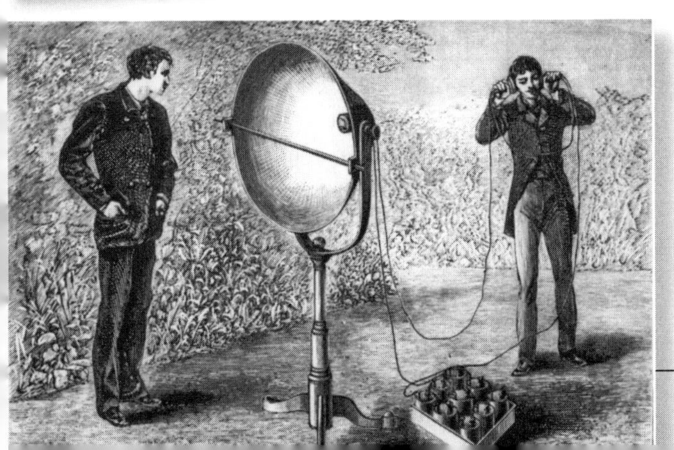

Charles Townes said he got chills when he later discovered that the park bench where he had his nocturnal breakthrough was located right in front of a house once owned by Alexander Graham Bell. In addition to his work on the telephone, Bell anticipated the use of light waves for communications with his photophone, patented in 1880.

1961

FRUSTRATED FASHION DESIGNER

A schoolboy dream deferred

Stoneham, Massachusetts, high school student Russell Colley astonished his teachers and his parents when he announced that his dream was to become a fashion designer. This was in the middle of the Roaring Twenties, but it certainly wasn't going to be "Anything Goes," as far as his family was concerned. He was told in no uncertain terms to direct his aspirations elsewhere. He showed an aptitude for engineering, and in 1928 went to work for B. F. Goodrich. Colley embraced his engineering career and left fashion behind.

It's funny how things work out, though.

In the 1930s Colley was called on to design a pressurized suit for pilot Wiley Post, who then set a new altitude record of forty thousand feet. He continued to work on pressurized suits for navy pilots through and after World War II.

When Americans were ready to ride rockets into space, who better to make their space suits than Russell Colley? He personally measured each of the Mercury Seven astronauts for the custom-fit space suits they would wear on their forays into the great beyond.

The boy who wanted to be a fashion designer wound up, in the words of legendary *New York Times* obituary writer Robert McG. Thomas, "The Calvin Klein of space suit design."

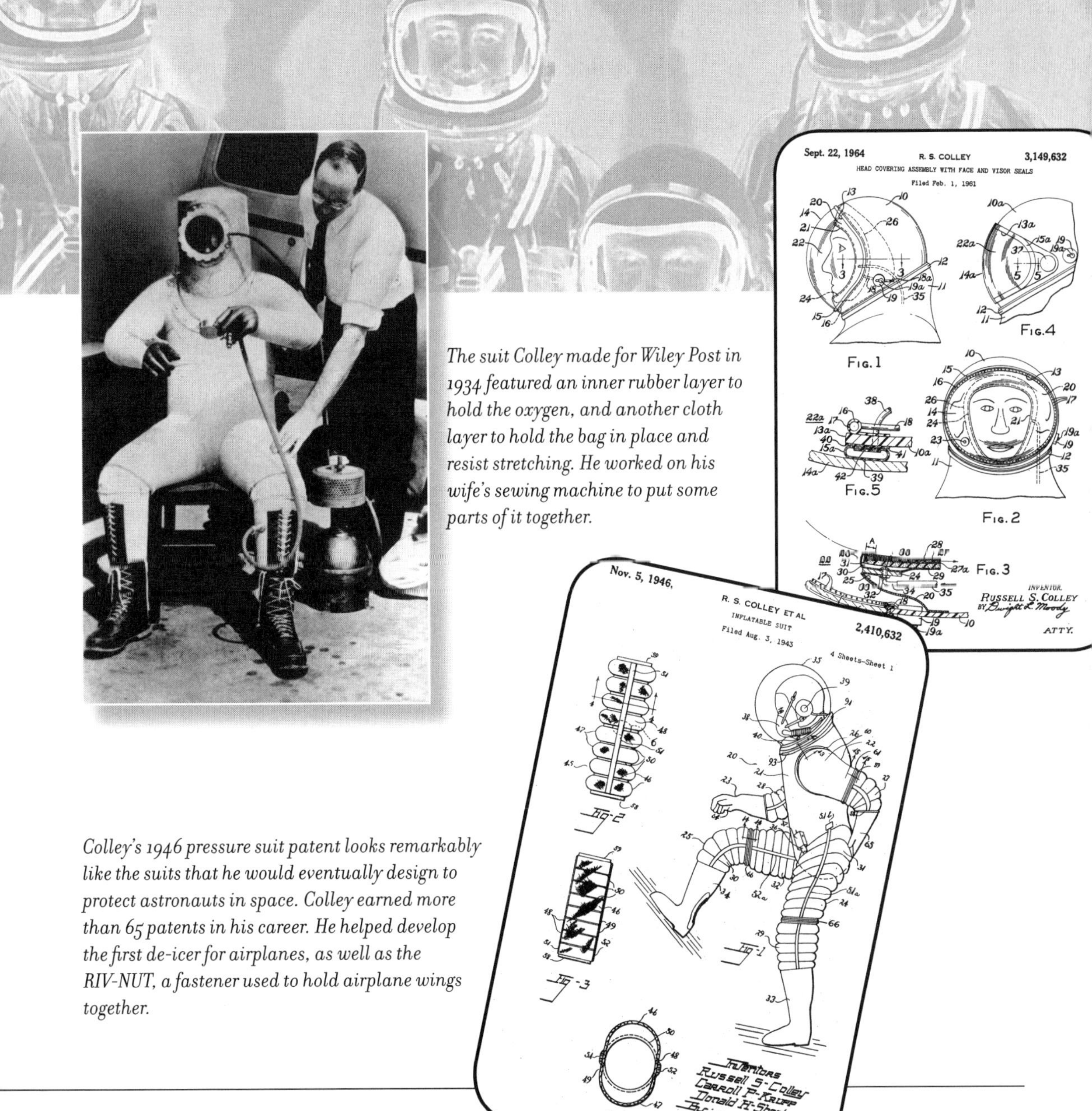

The suit Colley made for Wiley Post in 1934 featured an inner rubber layer to hold the oxygen, and another cloth layer to hold the bag in place and resist stretching. He worked on his wife's sewing machine to put some parts of it together.

Colley's 1946 pressure suit patent looks remarkably like the suits that he would eventually design to protect astronauts in space. Colley earned more than 65 patents in his career. He helped develop the first de-icer for airplanes, as well as the RIV-NUT, a fastener used to hold airplane wings together.

1963

HEART OF A TIGGER

Designed by a ventriloquist who was no dummy

If you know the name Paul Winchell, you probably know him as a famed ventriloquist, a children's TV host, or the original voice of Tigger in the Winnie the Pooh cartoons.

He is also the inventor of an artificial heart.

In the early 1960s, Winchell became friends with Dr. Henry Heimlich, eventually to earn fame for the Heimlich maneuver. Learning of Winchell's interest in medicine, Heimlich arranged for him to observe some surgeries. When Winchell observed a cardiac patient die during surgery one day, it gave him the idea of an implantable artificial heart. Encouraged by Dr. Heimlich, he designed and patented one in 1963. He offered his heart to the American Heart Association, but met with rejection.

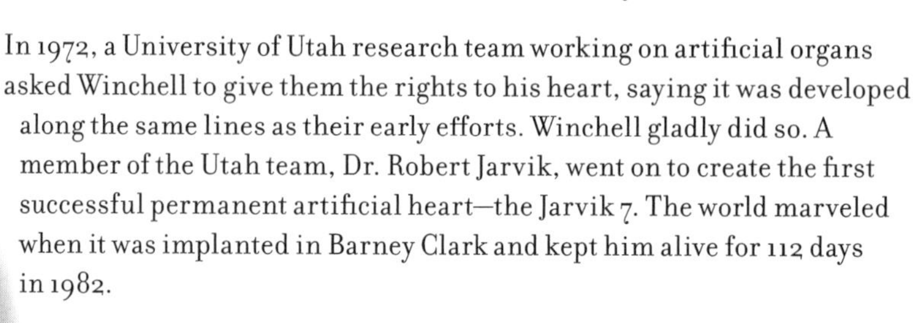

In 1972, a University of Utah research team working on artificial organs asked Winchell to give them the rights to his heart, saying it was developed along the same lines as their early efforts. Winchell gladly did so. A member of the Utah team, Dr. Robert Jarvik, went on to create the first successful permanent artificial heart—the Jarvik 7. The world marveled when it was implanted in Barney Clark and kept him alive for 112 days in 1982.

Did Winchell's heart play a role in the design of the Jarvik 7? Jarvik says that's nothing but a myth, that the two are fundamentally different. Dr. Heimlich has offered a different assessment. "The basic principle used in Paul Winchell's heart and the Jarvik 7 heart is exactly the same."

It's nice to imagine a little Tigger in every heart.

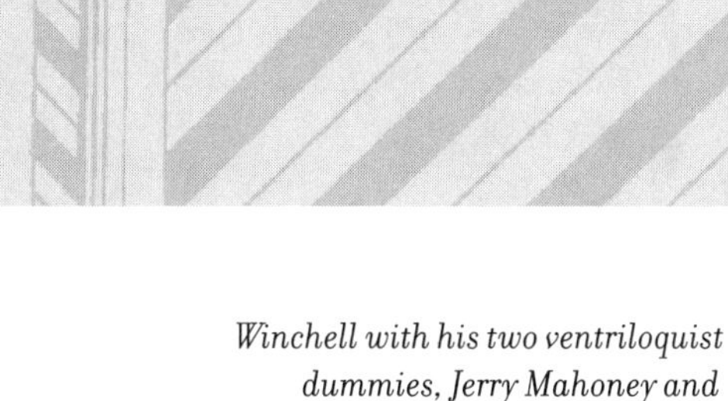

Winchell with his two ventriloquist dummies, Jerry Mahoney and Knucklehead Smiff.

July 16, 1963 P. WINCHELL 3,097,366
ARTIFICIAL HEART
4 Sheets-Sheet 2
Filed Feb. 6, 1961

INVENTOR.
PAUL WINCHELL
BY
ATTORNEY

Winchell had the bouncy enthusiasm of Tigger, patenting numerous inventions (including an invisible garter) and studying everything from acupuncture to astrology. In asking him to donate his patent rights, the Utah team promised him access to their facilities for a "joint effort in artificial heart research." Winchell always felt he was unfairly denied a share of the credit for their work.

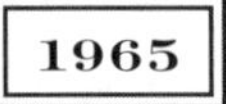

1965

FROM HISS TO BANG

What's that buzz? Tell me what's happening!

In 1960, NASA launched a giant Mylar balloon called Echo 1 into Earth orbit. It was the world's first primitive communications satellite. Bell Labs built a horn-shaped radio antenna on a hilltop in New Jersey to bounce signals off of it. After the Echo program was made obsolete by more elaborate communications satellites, two radio astronomers converted the antenna into a radio telescope. But Arno Penzias and Robert Wilson ran into a vexing problem.

There was some hiss in the system.

The faint background noise threatened to interfere with their delicate observations of deep space. So they set out to eliminate it. They kicked out some pigeons nesting in the horn, and swept out their droppings. They conducted elaborate diagnostic tests. They took the telescope apart, machined all the moving parts, and put it back together again. Nothing made the hiss go away. They became obsessed with trying to figure it out.

Then they heard about a scientist at Princeton who was saying that if the "Big Bang" theory of how the universe was created was true, there should be some low-level background radiation left over—just the hiss that Penzias and Wilson were getting.

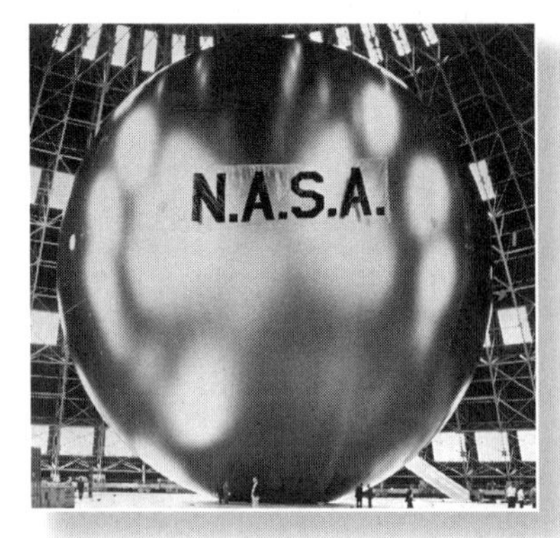

Largely forgotten today, the hundred-foot Mylar balloons of the Echo program were giant reflecting mirrors. They were sometimes known as "satelloons." Only two were launched into space. They were superseded in 1962 by Telstar, the first electronic communications satellite.

What they thought was a problem with their telescope turned out to be powerful proof that the universe began with a bang. Today it is called cosmic background radiation. And its accidental discovery earned Penzias and Wilson the Nobel Prize.

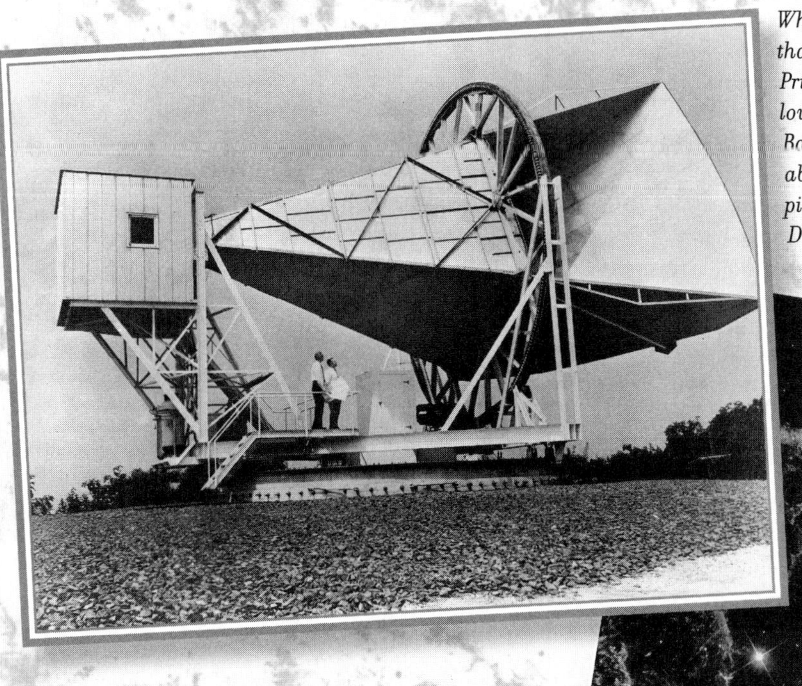

When Penzias and Wilson discovered that scientist Robert Dickie at Princeton had a theory about low-level radiation from the Big Bang, they called him to tell him about the hiss their telescope was picking up. "We've been scooped," Dickie told his fellow researchers.

1965

PATENTLY ABSURD II

More of the strange from birth to (almost) death

In 1965, George and Charlotte Blonsky were looking for a way to make childbirth easier, but the idea they came up with is a bit terrifying to contemplate. The expectant mother lies on a turntable that can be spun around at a speed of up to 90 rpm, achieving a force of seven times gravity, thus propelling the baby into a strategically placed net. The patent claimed that this apparatus would help civilized women give birth with less stress!

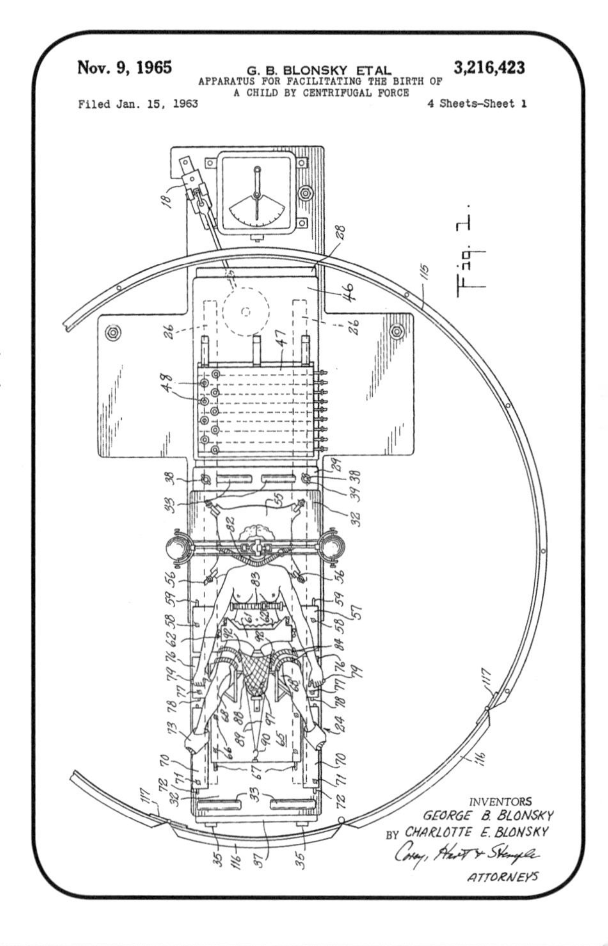

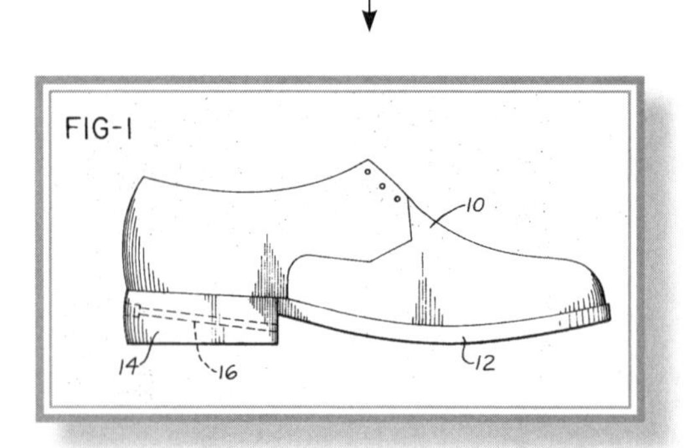

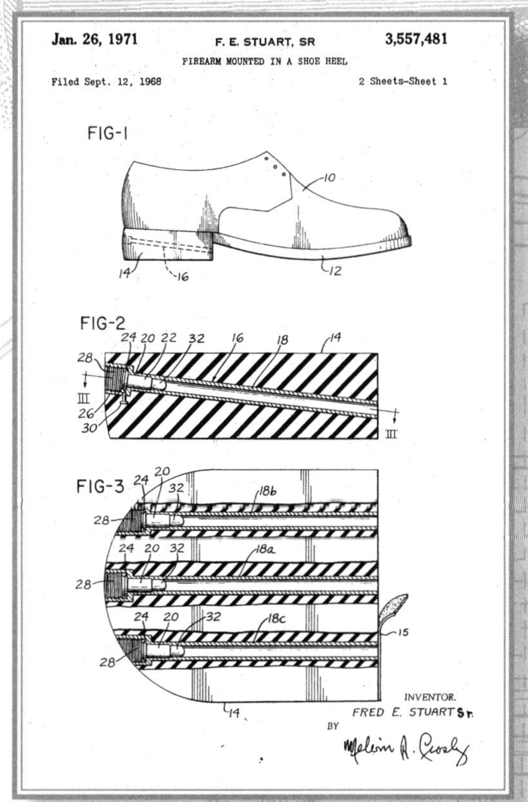

In 1971, F. E. Stuart patented a firearm mounted in a shoe heel, designed to be used by "authorized personnel" under "extraordinary circumstances." A device capable of giving new meaning to the expression "These heels are killing me!" Probably NOT the device for Agent 86, Maxwell Smart, who might get confused while answering his shoe phone, with tragic results.

The "Annunciator for the Supposed Dead" was a device dreamed up by William White of Topeka, Kansas, in 1891. It was designed for the not-quite-deceased person who accidentally ended up in a coffin. It allowed the undead to push a lever that rang an alarm in the cemetery office. If they came to after hours, however, they would just have to rest in peace . . . at least until morning.

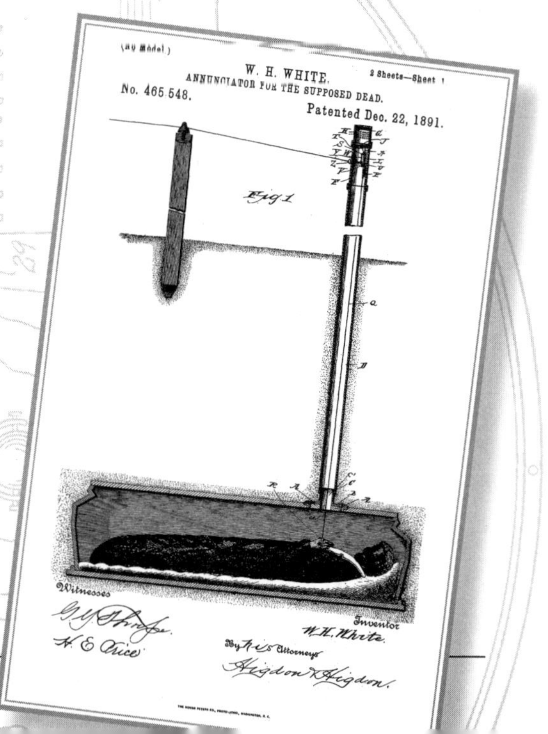

1965

THE KERNEL QUEST

The corniest story in this book

He was the first to admit it: he was a "funny-looking farmer with a funny-sounding name." He graduated with a degree in agronomy from Purdue in 1928, worked as a county agricultural agent, and managed a twelve-thousand-acre farm. He also spent time as a high school teacher, and a fertilizer executive.

Somewhere along the way, this bow-tie-wearing farmer became interested in breeding a new variety of corn. He hooked up with a partner named Charlie Bowman, and they spent decades crossbreeding more than thirty thousand corn hybrids. In 1965, they finally came up with the perfect kernel.

For popping, that is.

They named their new popcorn Red Bow. It was twice as expensive to produce as regular popping corn, and they couldn't get any big company to pick it up. So they started small, selling it out of the back of a station wagon to stores across Indiana. Real success eluded them until the fateful day they consulted with a Chicago marketing executive who charged them $13,000 for the following advice:

Name the popcorn after the funny-looking farmer.

"They came up with the same name my mother did for free," laughed Orville Redenbacher years later. So Red Bow turned into Orville Redenbacher's Gourmet Popping Corn, and quickly became the most popular popcorn in the land.

Redenbacher played the Sousaphone in the Purdue band. He was in his sixties when his popcorn hit the big time. Its success was due in no small part to his abilities as a spokesman for the product. He appeared in numerous ads, and, more than a decade after his death, remains a popular pitchman for the product he developed.

"My Orville Redenbacher's® Gourmet™ Popping Corn will bowl you over!"

"It pops lighter, fluffier, and a whole lot more."
Orville Redenbacher

4oz. of mine

4oz. of ordinary

Popcorn hadn't changed much in five thousand years until Redenbacher and Bowman came along. Most kinds of popping corn expand to about twenty times their size when popped. The Redenbacher-Bowman variety expands to about forty times its volume, making it lighter and fluffier. It also leaves fewer unpopped kernels, which Redenbacher called "the shy ones."

1968

FUTURE SHOCK

The block of wood that revolutionized computing

On December 9, 1968, Doug Engelbart unveiled his revolutionary new invention: "The X-Y Position Indicator for a Display System." Or, as he and his team liked to refer to their new wooden device, "the mouse."

For years, Engelbart had been thinking about how to make computers adapt to people, instead of the other way around. He and his team at the Stanford Research Institute had been working on ideas to do just that, and they presented them at a San Francisco computing conference, in a demonstration that was astonishingly ahead of its time.

In a day before e-mail or the Internet, when entire companies might have just one computer (or none), Engelbart asked his audience to consider the farfetched notion of a computer on each person's desk, responsive to every need. Then he grabbed his prototype mouse and took the crowd on a tour of the future. He showed them such radical things as cutting and pasting, folders, hypertext, using different windows on the screen, even working jointly with a remote user—all of which seemed about as fantastic as a *Star Trek* phaser. People in the audience actually started climbing on stage to see him use the mouse to move around what he called a "tracking spot" on the screen.

That demonstration is sometimes referred to as "the mother of all demos." It not only introduced the now-ubiquitous mouse—it offered a 20/20 vision of a revolution to come.

> "I don't know why we call it a mouse. It started that way and we never changed it."
>
> —Douglas Engelbart, December 9, 1968

Engelbart's mouse was a block of wood with two wheels, one for up-and-down motion, the other for back-and-forth. It had three buttons on it. It worked in conjunction with a controller mounted to the left of his keyboard, which featured five special-function keys.

The way we interface with the computer might be very different today if Engelbart's team had gone with another idea: using a knee brace to control the cursor on the screen.

The first mouse available to the public was developed by Xerox in the late 1970s, and cost $400. In the 1980s, Steve Jobs tasked a team of designers with the job of creating a mouse for Apple that would cost one-tenth that price. The team, led by Dan Hovey, built their first prototype from parts that included the roller ball from some Ban deodorant and a plastic butter dish.

1969

FOR THE RECORD

The right way to remember a rocketeer

On January 17, 1969, the *New York Times* ran a correction on an article published in the paper forty-nine years earlier. The original article, which appeared in 1920, mocked rocket pioneer Robert Goddard for suggesting that it would be possible to launch a rocket to the moon. The article asserted that a rocket could not operate in a vacuum, and belittled Goddard's understanding of basic science.

> That Professor Goddard, with his "chair" in Clark College, and the countenancing of the Smithsonian Institution, does not know the relation of action to reaction and the need to have something better than a vacuum against which to react—to say that would be absurd. Of course he only seems to lack the knowledge ladled out daily in high schools.

Goddard survived this assault on his scientific acumen, going on to launch the world's first liquid-powered rocket in 1926. His pioneering work helped pave the way for the space program of the 1960s. The correction appeared in the paper the day after the launch of *Apollo 11*, which would land a man on the moon three days later.

"It is now definitely established that a rocket can function in a vacuum," the correction announced, with admirable understatement.

"The *Times* regrets the error."

R. H. GODDARD.
ROCKET APPARATUS.
APPLICATION FILED DEC. 23, 1915.
1,194,496. Patented Aug. 15, 1916.

Goddard patented most of the basic features of a modern rocket, and after his death his family was awarded a million dollars by NASA for their use.

Goddard's quest to build a space rocket began on October 19, 1899. The seventeen-year-old boy had climbed high in the branches of a cherry tree at his family's farm in Worcester, Massachusetts. "I imagined how wonderful it would be to make some device which had even the possibility of going to Mars," he later wrote. "I was a very different boy when I descended the tree." From then on, Goddard referred to October 19 as "Anniversary Day," and would try to visit the tree each year on that day.

Goddard conducted his rocket launches on a farm in Auburn, Massachusetts, until a rocket exploded in 1929. Alarmed neighbors called the police, and the state fire marshal told Goddard he had to take his rockets elsewhere. Charles Lindbergh helped him secure funding to build a launch site in the empty expanses of the West. The spot he selected: Roswell, New Mexico.

1969

LO AND BEHOLD

The secret birth of a global marvel

It was the year of Woodstock and of the first moon landing. But the most earthshaking event of the year may have been one that hardly anyone noticed. It involved a refrigerator-sized computer that sits today in a third-floor storage space at UCLA.

The first computer on the Internet.

The U.S. government was funding scientific research through the Advanced Research Project Agency, better known as ARPA. But there was no way to share information between giant computers working at different sites. So the government decided to connect them on a new network called ARPANET.

The first node on the new network was a switching computer called an Interface Message Processor. It was so big it had to be hoisted to its third-floor location on a forklift. It was hooked up to UCLA's mainframe computer on September 2, 1969. Internet pioneer Len Kleinrock, who supervised the installation of the machine, says: "The Internet took its first breath of life that day."

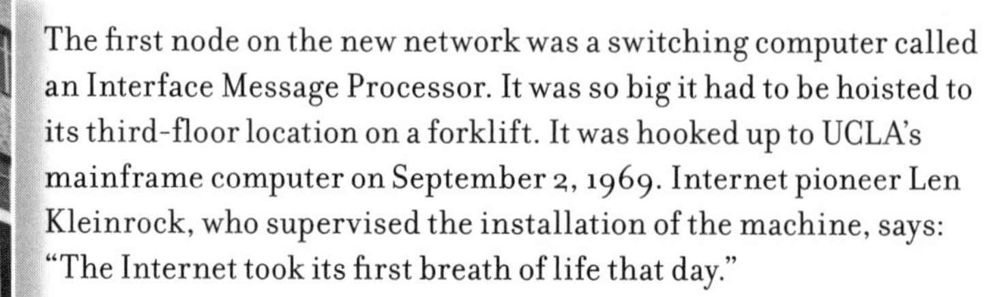

Less than two months later, technicians had that computer talking over phone lines with another one at Stanford. By 1971, there were fifteen computers wired together on the fledgling network that would eventually evolve into the global network we all depend on today.

The Interface Message Processor was so big it had to be lifted to the third floor by a forklift and eased in through a window. It is still there.

Kleinrock relishes describing the moment on October 29, 1969, that UCLA programmers tried for the first time to log on to the computer at Stanford:

> *Charlie Klein with a headset and a microphone, connected over this network, with a voice line to the fellow at the other end. Charlie typed the L, said, "Did you get the L?" The other guy said, "I got the L." He typed the o. "Did you get the o?" "I got the o." He typed in the g. Crash. So the first message on the Internet was lo . . . as in "Lo and behold."*

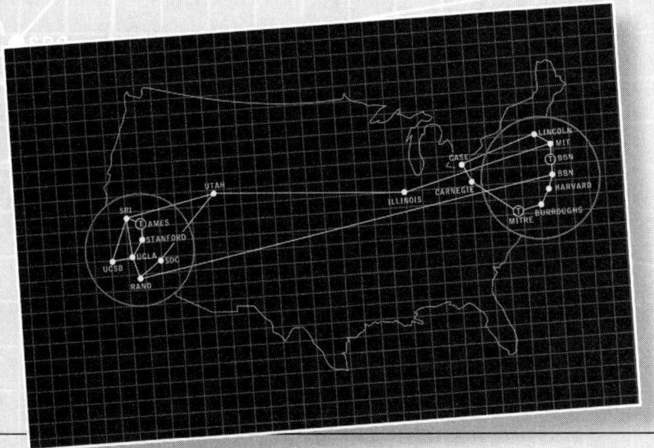

Kleinrock, who helped develop some of the packet-switching ideas that led to the Internet, says that in the early days of ARPANET, "Nobody wanted to join this network, A, and B, when I did my research, nobody believed it would work."

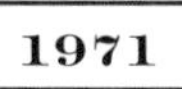

NOTE TO SELF

The birth of e-mail

In 1971, a computer programmer named Ray Tomlinson spent a few hours playing around on a couple of computers at work and came up with a little something we call e-mail.

He sent the first one to himself.

There were about fifteen computers hooked up on ARPANET, the predecessor to the Internet, but no easy way to direct a message to an individual on another computer. Tomlinson thought that would be a neat feature to have at his disposal, and he set out to create it.

He worked with two machines about fifteen feet apart, sending test messages from one and then going into his mailbox on the second machine. Eventually, one of his messages got through, and e-mail was born.

"It probably took four, five, six hours to do, spread over a week or two," says Tomlinson. "I was not working full-time on it. I had real work to do, too."

Tomlinson selected a little-used symbol above the 2 on his keyboard to specify to which computer the message was headed. Today the @ symbol is the ubiquitous icon of e-mail, and Ray Tomlinson's invention connects people all over the globe.

Not bad for a few hours of fooling around.

Ray Tomlinson isn't exactly a household name. But he has gotten a fair amount of attention. When a question about him popped up on the game show Jeopardy!, *his mother, a faithful fan of the program, was tickled. She said she was happy to finally see a question she knew the answer to.*

"IT WAS A HACK— A NEAT THING TO TRY OUT."

—RAY TOMLINSON ON THE INVENTION OF E-MAIL

E-mail spam was born on May 3, 1978, when a marketer at Digital Equipment Corporation (DEC) sent out an e-mail to every single ARPANET user on the West Coast, trumpeting the company's latest computer. The message went to—gasp—nearly four hundred people. Today it is estimated that 200 billion spam e-mails are sent every twenty-four hours.

The @ symbol has been on typewriters for more than a hundred years, used in bookkeeping to signify price per unit. Some scholars suggest it comes from the French word for "at," which is à. Italian scholar Giorgio Stabile has uncovered what he says is the first known use of the symbol in a 1536 letter from a Florentine merchant named Francesco Lapi. Stabile says that it represents an amphora, *a measure of volume based on a large jar.*

1977

STICKING POWER

It took plenty of stick-to-itiveness to make this product come to life

In 1964, Spencer Silver invented a glue that didn't work very well. It wasn't particularly sticky, and it wouldn't dry. You couldn't really attach anything together with it. Still, the 3M chemist thought it must have some use. For years, he would visit anyone at the 3M headquarters in St. Paul, Minnesota, who would listen, and show them his goop, encouraging them to think of a use for it.

Nearly a decade later, another 3M employee, named Art Fry, was singing in the choir at a St. Paul Church. He liked to mark the songs in his hymnal with slips of paper, but sometimes when he flipped through the pages, the paper fell out. Then Fry remembered a conversation he had with Spencer Silver a couple of years before. By coating the paper slips with Silver's stuff, he could make a bookmark that wouldn't fall out.

So was born the Post-it note.

The product was a big hit at 3M headquarters, but it failed miserably when first test-marketed in 1977. Nobody wanted to pay a dollar for a pad of sticky notepaper they couldn't figure out how to use. So 3M hit on the idea of giving the pads away and hoping people would get hooked. They blanketed Boise, Idaho, with free samples in what became known in 3M lore as the "Boise Blitz."

It worked—and Post-it notes have stuck with us ever since.

Post-it®

Spencer Silver's first idea for the glue was to coat a bulletin board with it, so you could stick papers to it.

What makes Spencer Silver's adhesive stick a little, but not a lot? The scientific answer is that acrylate copolymers form a suspension of microspheres, which gives the adhesive a pebbled surface that has less stickiness than a flat surface. Aren't you glad you asked?

Post-it®
Notes
Notas
654
3 in/po x 3 in/po
76,2 mm x 76,2 mm
Cont. 100 hojas
3M

1975

CELEBRITY PATENTS

From Zeppo Marx to Mark Twain

Sometimes, patents make little-known people famous. Other times, famous people make little-known patents. Here are a few.

Legendary boxer Jack Johnson, the first black heavyweight champion, was granted a patent in 1922 for an adjustable wrench. Presumably, not for use in the ring.

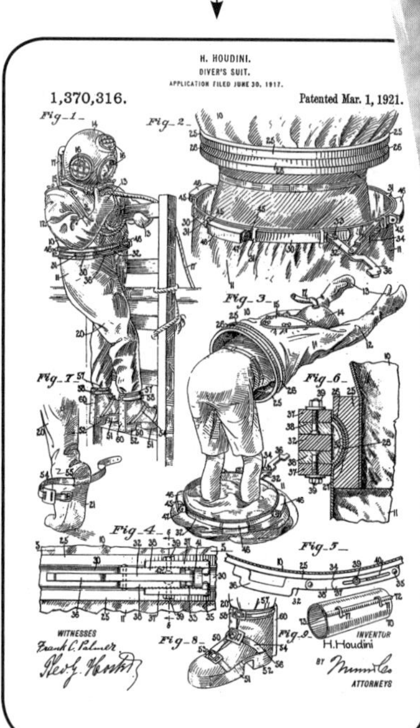

Escape artist and magician Harry Houdini obtained a patent in 1921 for a diving apparatus. It should come as no surprise that Houdini's suit was designed "to permit the diver . . . to quickly divest himself of the suit while being submerged and to safely escape to the surface."

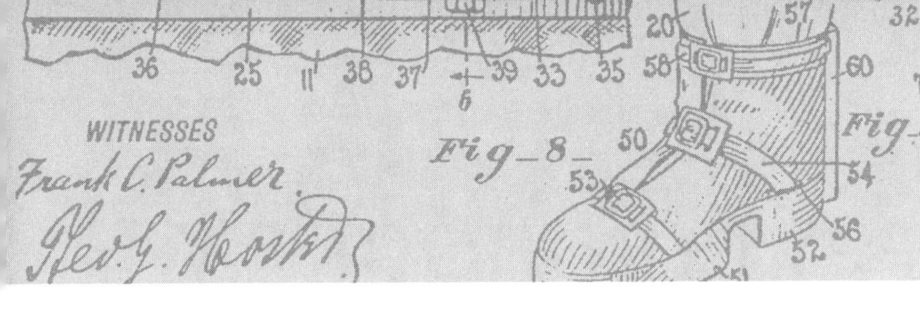

Zeppo Marx (he's the one at the bottom) was the straight man in the Marx Brothers movies. In 1967, Zeppo (did you know his real name was Herbert?) coinvented a special wristwatch for cardiac patients. It has two dials, one driven by the wearer's pulse and another operating at a rate corresponding to a normal heartbeat. If the pulse-driven watch started running fast, the patient would know to slow down.

Author Mark Twain was issued three patents. One was for a scrapbook with adhesive-coated pages. It was marketed as "Mark Twain's Patent Scrapbook" (oh the clever thinking that went into that name!) and became a moderate success. Another was for a complicated game designed to help children remember history. (Also the purpose of books like this one, come to think about it.)

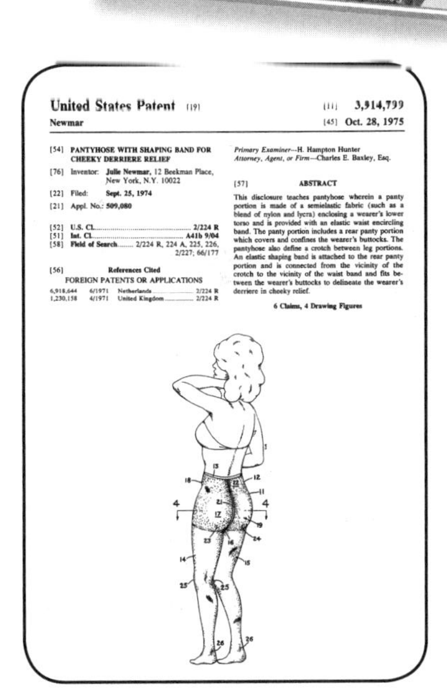

United States Patent [19] **Newmar**

[11] **3,914,799**
[45] **Oct. 28, 1975**

[54] **PANTYHOSE WITH SHAPING BAND FOR CHEEKY DERRIERE RELIEF**

[76] Inventor: **Julie Newmar,** 12 Beekman Place, New York, N.Y. 10022

[22] Filed: **Sept. 25, 1974**

[21] Appl. No.: **509,080**

[52] **U.S. Cl.** .... **2/224 R**
[51] **Int. Cl.** **A41b 9/04**
[58] **Field of Search** .... 2/224 R, 224 A, 225, 226, 2/227; 66/177

[56] **References Cited**
FOREIGN PATENTS OR APPLICATIONS

6,918,644 6/1971 Netherlands .... 2/224 R
1,230,158 4/1971 United Kingdom .... 2/224 R

Primary Examiner—H. Hampton Hunter
Attorney, Agent, or Firm—Charles E. Baxley, Esq.

[57] **ABSTRACT**

This disclosure teaches pantyhose wherein a panty portion is made of a semielastic fabric (such as a blend of nylon and lycra) enclosing a wearer's lower torso and is provided with an elastic waist encircling band. The panty portion includes a rear panty portion which covers and confines the wearer's buttocks. The pantyhose also define a crotch between leg portions. An elastic shaping band is attached to the rear panty portion and is connected from the vicinity of the crotch to the vicinity of the waist band and fits between the wearer's buttocks to delineate the wearer's derriere in cheeky relief.

6 Claims, 4 Drawing Figures

Last, but not least, actress Julie Newmar, still remembered as the original Cat Woman of the Batman *TV series, patented "Nudemar" pantyhose in 1975. According to an article in* People *magazine, Newmar said her invention could "make your derriere look like an apple instead of a ham sandwich." She also received a patent for a brassiere designed for older women.*

1978

THE HAPPY NON-HOOKER

A curve ball on the golf course

Fred Holmstrom wasn't a golfer. But the California physicist became interested in golf balls when he read an article about their aerodynamic properties. That's when he and his friend Dan Nepela decided to do something that would rock the golf world and spawn a million-dollar lawsuit.

They invented a golf ball that always flew straight.

Winston Churchill described golf as "a game whose aim is to hit a very small ball into an even smaller hole, with weapons singularly ill-designed for the purpose." Every duffer whose flawed swing has ruined a Saturday morning can appreciate the value of a ball that can't be hooked or sliced.

And it wasn't hard to make one. Holmstrom and Nepela created a ball with shallower dimples on the poles, and regular dimples around the equator. It flew like it had a gyroscope on board. They called it the Polara and sold hundreds of thousands of them.

To be used in competition, golf balls had to pass four tests at the U.S. Golf Association. The Polara passed every one, but the USGA banned it anyway, saying it wasn't in the spirit of the game. They actually changed their rules to make it illegal. The bad publicity put the Polara out of business. Holmstrom and Nepela sued, eventually settling out of court.

And golfers were forced to improve the hard way—with practice.

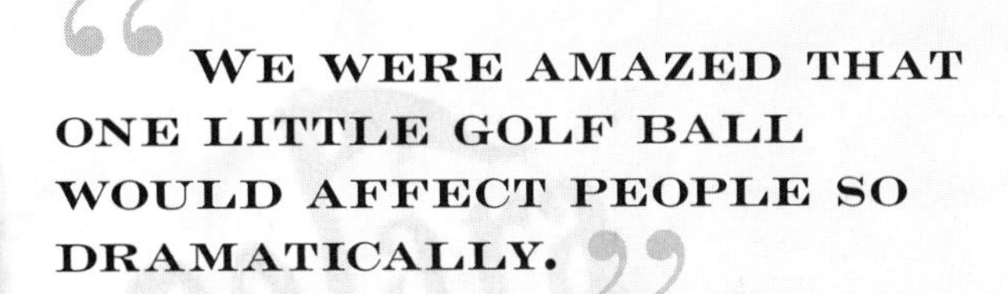

—FRED HOLMSTROM

In their first experiment, Holmstrom and Nepela took dozens of golf balls, and filled in the dimples around the poles with white putty. Then they went to the golf course with a guinea-pig golfer. They asked him to hit a slice. The ball started to curve, then straightened itself out. The golfer dropped the club in surprise: "I can't believe what I just saw."

The ball that always flew straight was off the market for more than twenty-five years, but it has recently had a rebirth. Pounce Sports purchased the rights to the ball and reintroduced it in 2008. But it is still against the rules to use it in competition.

1997

THREE MATHEMATICIANS

Searching for the story behind the search

Once upon a time, there were three mathematicians. In the late 1800s, Russian mathematician Andrei Markov did groundbreaking work in probability theory. In 1938, American mathematician Edward Kasner asked his nine-year-old nephew to coin a name for a really big number. And in 1979, a mathematician named Michael Brin emigrated from Moscow to the United States.

That's the backstory of an idea that came to be called "BackRub," one of the most successful business/technology innovations of the twentieth century.

Never heard of it? Sure you have.

In the 1990s, a Stanford graduate student named Larry Page used Markov's theories to map the World Wide Web. He eventually partnered with a fellow grad student, Michael Brin's son, Sergey, to create "BackRub." It ranked the popularity of Web pages based on their back-links, that is, the links from other Web sites to that page.

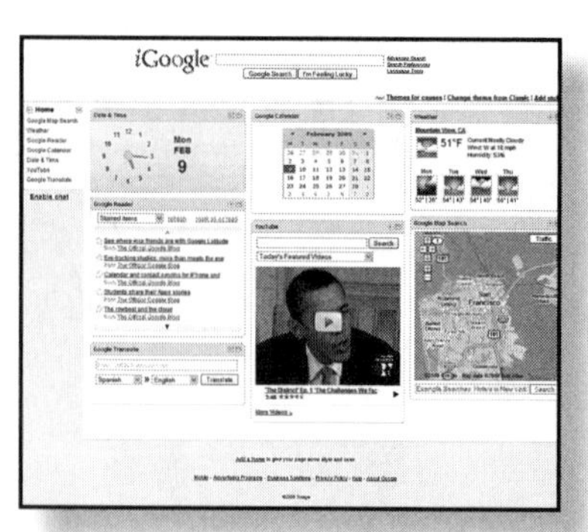

In the fall of 1997, they were ready to present BackRub to the world, but decided that it needed a better name. Since it involved organizing so much data, one idea was to name it after a really big number. That's where Edward Kasner's nine-year-old nephew, Milton Sirotta, came in. Asked by his uncle back in 1938 to suggest

The patent behind Google has the nonsexy title "Method for Node-ranking in a Linked Database."

a new name for a number consisting of 1 followed by a hundred zeroes, he came up with the word "googol."

Turned out Larry Page and Sergey Brin couldn't spell very well. Then again, they couldn't do what we can do now:

Google it.

Google has made Brin and Page two of the wealthiest people on Earth. In 2009, Forbes *magazine ranked them each as tied for being the twenty-sixth-richest person in the world, estimating that each is worth 12 billion dollars.*

Google's headquarters is called Googleplex, which derives from yet another number. Milton Sirotta, mathematician Kasner's young nephew, suggested that googolplex should be the number one, "followed by writing zeroes until you get tired." Kasner defined it as 10 to the power of googol, or 10^{google}. It is a number so large that there isn't enough space in the universe to write it down.

SPACE 2.0

Moving to the future at warp speed

On October 4, 1957, the launch of *Sputnik* gave birth to the space age. Forty-seven years later to the day, another launch paved the way for a revolution in space travel.

That's the day that Burt Rutan's *SpaceShipOne* won the $10 million Ansari X-Prize for becoming the first privately funded spacecraft to make two journeys into space within two weeks. The X-prize was modeled on the Orteig Prize, which inspired Lindbergh's flight to Paris.

Rutan's spacecraft was the size of a small biplane with a rocket engine that ran on laughing gas and rubber. Carried aloft and launched from a cargo plane called the *White Knight*, it carried a single pilot on a suborbital flight more than seventy miles into space before gliding back to earth.

That flight marked the beginning of a new space race, between businesses instead of governments. Visionaries are now working feverishly on space taxies, space hotels, and a new generation of space liners.

The final frontier draws ever closer.

A futuristic view of Spaceport America, established by the state of New Mexico, which hopes to become America's launching pad of the future. More than seven private spaceports were also under construction by late 2008.

Burt Rutan is building a spaceship for Virgin Atlantic, designed to carry a pilot and five passengers eighty miles into the sky, where they will experience four minutes of weightlessness, then hit six G's as they glide back down to earth. The anticipated ticket price for a ride on SpaceShipTwo: $200,000 per person.

In the first half-century of space travel, about five hundred people have made it into space. Flamboyant Virgin founder Richard Branson hopes to take fifty thousand up in the next ten years. He wants to go on the first trip himself—with his mother and father.

SELECTED SOURCES

Information is only as good as its source. Readers deserve to know the principal sources for each story, so that they can judge for themselves how accurate it is, and where to go to find out more. I've provided a story-by-story list with the selected sources for each.

When I began looking for stories to include in this book, I discovered several terrific Web sites filled with stories about science and invention. At the Lemelson-MIT Inventor of the Week Archive (web.mit.edu/invent/i-archive.html), you can browse through stories of hundreds of inventors. The Engines of Our Ingenuity site (www.uh.edu/engines/) contains thousands of scripts of the same-named radio series written and hosted by John Lienhard, who is the author of a growing number of books about inventing. These make a great jumping-off point for anyone interested in this subject.

There are a handful of online sources that I consulted so often that they deserve special mention. The Encyclopaedia Britannica (www.britannica.com) is a wonderful source for basic historical information. Wikipedia (www.wikipedia.com) proves more useful and accurate every year, although it is a good idea to double-check anything found there. The *New York Times* archive, available through many libraries, allows word-searching of stories going back to 1857. One of the most fun weeks I spent on this project was perusing the science-oriented obituaries of the twentieth century, where I found more than a dozen of the stories that made it into the book. The *Time* magazine archive (www.time.com) allows word-searching of articles back to 1923. *American Heritage* and *American Heritage Invention & Technology* (www.americanheritage.com) have thousands of well-researched history articles just fingertips away. I went back to each of these sources many, many times. I also made extensive use of Google Book Search (books.google.com) to instantly put my hands on texts that otherwise might have taken weeks to locate.

Bath Time: *Great Accidents in Science that Changed the World* by Jerome Sydney Meyer; *Archimedes and the Roman Imagination* by Mary Jaeger.

Persian Power: *Ancient Inventions* by Peter James and Nick Thorpe; "Riddle of Baghdad's Batteries," BBC online, news.bbc.co.uk/2/hi/science/nature/2804257.stm.

Ancient Hero: *Ancient Inventions* by Peter James and Nick Thorpe; *Encyclopedia Americana.*

Shake, Rattle, and Roll: *Ancient Chinese Inventions* by Yinke Deng.

Easy as 1, 2, 3: *The History of Mathematics* by Dave Burton; *A History of Mathematics* by Florian Cajori.

Flight Before Wright: "Eilmer of Malmesbury, an Eleventh Century Aviator," by Lynn White Jr., *Technology and Culture*, Vol. 2, No. 2 (Spring 1961), pp. 97-111, accessed via JSTOR, www.jstor.org/stable/3101411; *Like Sex with Gods* by Bayla Singer.

Lippershey's Looker: "On the First Invention of the Telescopes" by Dr. G. Moll, *The Journal of the Royal Institution of Great Britain*, vol 1, 1831.

The Alchemist: *Newton* by Peter Ackroyd; *In the Presence of the Creator* by Gale Christianson.

Sybilla Masters: *Encyclopedia of American Women in Business* by Carol Krismann; *The Pennsylvania Magazine of History and Biography*, vol. viii, 1884.

Paper Trail: *They All Laughed*, by Ira Flatow; *Get It on Paper*, a documentary written and produced for The History Channel® by Kate Raisz, with Rick Beyer as executive producer.

Ben the Weatherman: *Benjamin Franklin* by Walter Isaacson; *The Turbulent History of Weather Prediction from Franklin's Kite to El Niño* by John D. Cox.

Ring of Fire: *The Victorian Internet*, by Tom Standage; *A Wired World*, a documentary written and produced for The History Channel® by Rick Beyer.

First Car: *World History of the Automobile* by Erick Eckerman; *Encyclopedia of Earth.* It was my pleasure to inspect one of Cugnot's first cars at the Musée des arts et métiers in Paris in 2008.

The Music Man: *The Neptune File* by Tom Standage.

The Mechanical Internet: *The Victorian Internet*, by Tom Standage; *The Early History of Data Networks* by Gerard J. Holzmann and Björn Pehrson; my tour of restored Chappe telegraph station in Saverne, France.

The Other Ben: *Memoir of Sir Benjamin Thompson, Count Rumford* by George Edward Ellis; *Benjamin Thompson, Count Rumford* by Sanford C. Brown; *American Science and Invention* by Mitchell Wilson; "The Strange Forgotten Life of America's Other Ben Franklin" by Nicholas Delbanco, *American Heritage*, September 1993.

Shell Shock: *Harper Encyclopedia of Military Biography*; "Honour for the Man Who Changed the Face of War" by June Southworth, *London Daily Mail*, July 28, 1994.

First Car, USA Edition: "Was This America's First Steamboat, Locomotive, and Car?" by Steven Lubar, *American Heritage Invention & Technology*, Spring 2006; *They Made America* by Harold Evans.

Can Do!: *Food and Culture: A Reader* by Carole Counihan; *Circles* by James Burke.

Bah-bump Goes the Stethoscope: *Great Adventures in Medicine*, edited by Samuel Rapport and

Helen Wright; "The Chance Invention That Changed Medicine," by Dr. John G. Leyden, *Saturday Evening Post*, May 2001; "The Inventor of the Stethoscope: René Laennec," by Harry Bloch, *Journal of Family Practice*, August 1993.

First Computer: *Bit by Bit* by Stan Augarten; *The Bride of Science* by Benjamin Wooley.

Picture This: *The Origins of Photography* by Hemut Gernsheim.

A Natural Selection: *The Life and Letters of Charles Darwin*, edited by his son Francis Darwin; *Darwin and the* Beagle by A. Moorehead.

The Long Night: *Men of Mathematics* by E. T. Bell; *Asimov's Chronology of Science and Discovery* by Isaac Asimov.

Of Railroads and Trumpets: *Sounds of Our Times* by Robert T. Beyer; *Archimedes to Hawking* by Clifford A. Pickover.

The Elastic Man: *Good Morning*, a documentary produced by Ron Blau for The History Channel® with Rick Beyer as executive producer; *Gum Elastic* by Charles Goodyear.

Patent President: *Lincoln: The Prairie Years* by Carl Sandburg; *Herndon's Lincoln* by William H. Herndon; additional information supplied by Harry Rubenstein at the Smithsonian's National Museum of American History and Selma Thomas of Watertown Productions.

Going Up: *Eureka: Great Inventions and How They Happened* by Richard Platt; *Otis: Giving Rise to the Modern City: A History of the Otis Elevator Company* by Jason Goodwin; Otis Elevator Web site www.otisworldwide.com/d31-timeline.html.

Bigo's Beets: *Microbe Hunters* by Paul De Kruif; *Germ Hunter* by Elaine Marie Alphin and Elaine Verstraete.

Dem Dry Bones: "Dusting Off America's First Dinosaur" by Richard Ryder, *American Heritage*, March 1988; *The Legacy of the Mastodon* by Keith Stewart Thomson.

Telephone Tale: *Sounds of Our Times*, by Robert Beyer; *Philip Reis, Inventor of the Telephone* by Sylvanus Thompson; *Alexander Graham Bell: The Life and Times of the Man Who Invented the Telephone* by Edwin S. Grosvenor and Morgan Wesson; letter from Elisha Gray to A. L. Hays (March 19, 1876), Western Union Collection, Library of Congress.

Birth of a Beverage: *Coca erythroxylon (Vin Mariani)* by Mariani & Company; *Cocaine* by Dominic Streatfield; *For God, Country, and Coca-Cola* by Mark Prendergrast; Coca-Cola Web site http://www.cocacola.com; "How Coca-Cola Obtains Its Coca," *New York Times*, July 1, 1988.

Radio Prophet: "Through the Air in 1866" by Dr. Malvin E. Ring, *American Heritage Invention & Technology*, Fall 2003; *Mahlon Loomis, Inventor of Radio* by Thomas Appleby; Library of Congress Web site www.loc.gov/exhibits/treasures/trr083.html.

Chester's Champions: *Chester: More Than Earmuffs* by Nancy Porter; *Farmington [Maine] Town Register 1902-3*; *A History of Farmington, Franklin County, Maine, from the Earliest* by Francis Gould Butler. There is a great deal of misinformation about Chester Carlson, some of it fabricated in the mid-1970s by a newspaper reporter (now a journalism teacher) named Christopher Corbett, who thought it would be amusing to make up stuff about Chester when the Maine legislature launched Chester Greenwood Day. "When we got done, his mother wouldn't have known him," bragged Corbett years later, in the December 1997 issue of *Yankee Magazine*.

The Devil's Rope: "Barbed Wire," a "Patent Files" segment for *This Week in History* produced for The History Channel® by Julie Rosenberg, with Rick Beyer as executive producer; "Rival Inventors Claimed Device for Barbed Wire," *New York Times*, March 7, 1926.

Sweet and Sour: *They All Laughed* by Ira Flatow; *Sweet and Low* by Rich Cohen; "Benjamin Eisenstadt, 89, a Sweetener of Lives," *New York Times*, April 10, 1996.

The Lightbulb Before Edison: *They All Laughed* by Ira Flatow; *Eureka!* by Carol Chapman, Ruth Holmes; "J. W. Starr: Cincinnati's Forgotten Genius" by Charles D. Wrege, *Cincinnati Historical Society Bulletin*, Summer 1976, accessed at library.cincymuseum.org/.

Dirty Dishing: "The Woman Who Invented the Dishwasher," J. M. Fenster, *American Heritage Invention & Technology*, Fall 1999.

Fade to Black: *The Missing Reel* by Christopher Rawlence; *Turning Points in Film History* by Andrew J. Rausch.

The Undertaker's Revenge: *A Wired World*, a documentary produced for The History Channel® by Rick Beyer; *100 Years of Telephone Switching* by Robert J. Chapuis and Amos E. Joel; various articles supplied by the LaPorte Indiana Historical Society.

Matchmakers: *A History of Delaware County, Pennsylvania, and Its People* by John Woolf Jordan; various newspaper articles supplied by the Delaware County (Pennsylvania) Historical Society; various articles in the *New York Times*.

The Man with Wheels in His Head: "The Big Wheel" by Patrick Meehan, *Hyde Park Historical Society Newsletter*, Spring 2000; *The Devil in the White City* by Erik Larson.

The Hookless Hooker: "The History of the Zipper?" by Robert Friedel, *American Heritage Invention & Technology*, Summer 1994.

Cutting Edge: *King C. Gillette: The Man and His Wonderful Shaving Device* by Russell Adams; "K. C. Gillette Dead; Made Safety Razor," *New York Times*, July 11, 1932.

Röntgen's Rays: *Discovery by Chance: Science and the Unexpected* by Mary Batten.

Patently Absurd I: Based on patents found in the U.S. Patent Office.

That Giant Sucking Sound: "The Vacuum Cleaner," by Curt Wholeber, *American Heritage*, Spring 2006; "Vacuum Cleaner," a "Patent Files" segment for *This Week in History* produced for The History Channel® by Julie Rosenberg, with Rick Beyer as executive producer.

Accidents Happen: *Inventions and Discoveries* by Valerie-Anne Giscard d'Estaing; "Escape: Because Accidents Happen," a documentary in the PBS *NOVA* series; *A History of the Auto Glass Industry* by Karl K. Alberti.

A Wire in Winter: "How the Wire Coat Hanger Got Invented" by Gary Mussell, great-grandson of the inventor, users.vcnet.com/garym/hanger/hanger.html; *Extraordinary Uses for Ordinary Things* by *Reader's Digest*; *Origin of Everyday Things* by Johnny Acton, Tania Adams, Matt Packer.

Ticket to Ride: "Einstein's Big Idea," *NOVA*, http://www.pbs.org/wgbh/nova/einstein/.

Sweet Caresse: *Caresse Crosby: From Black Sun to Roccasinbalda* by Anne Conover; *They Made America* by Harold Evans; *The Camoisy Queen: A Life of Caresse Crosby* by Linda Hamalian.

Forever Young: *The Man Who Changed How Boys and Toys Were Made* by Bruce Watson; "The Toys That Built America" by Henry Petroski, *American Heritage Invention & Technology*, Spring 1998.

The Amazing Doctor Abrams: "Abrams Reactions," *Time*, November 12, 1923; "The King of Quacks" by J. D. Haines, *Skeptical Inquirer*, May 2002.

Turning Point: *Florence Lawrence, the Biograph Girl* by Kelly R. Brown.

Josephine's Joy: "The BAND-AID: It's Not Just a Quick Cosmetic Fix" by Curt Wohleber, *American Heritage Invention & Technology*, Summer 2005; "Modern Bandages Owe a Big Debt to One Clumsy Wife" by Ellen Javernick, *Cappers*, April 2006.

Otto's Dream: *The Chemical Languages of the Nervous System* by Josef Donnerer and Fred Lembeck; *The War of the Soups and the Sparks* by Elliot S. Valenstein.

The Teen Who Invented Television: *The Boy Who Invented Television* by Paul Schatzkin.

Tee Time: *The Singular History of the Golf Tee* by I. R. Valenta; "Golf Tee," a "Patent Files" segment for *This Week in History* produced for The History Channel® by Julie Rosenberg, with Rick Beyer as executive producer.

Fish Out of Water: *Eureka!: Great Inventions and How They Happened* by Richard Platt; "Clarence Birdseye, Father of Frozen Food," www.birdseye.com.

The Honey Bee Boogie: *Bees* by Candace Savage; *The Eureka Moment: 100 Key Scientific Discoveries of the 20th Century* by Rupert Lee.

Hair Today: "Sol H. Goldberg, Hairpin Maker, 53" *New York Times*, January 6, 1940; Patents from the U.S. Patent Office.

Einstein's Refrigerator: "The Einstein Szilard Refrigerators" by Gene Dannen, *Scientific American*, January 1997; "Einstein the Inventor" by Thomas Hughes, *American Heritage Invention & Technology*, Winter 1991.

The Mold That Saved Millions: *In Search of Penicillin*, by David Wilson; *Alexander Fleming: The Man and Myth*, by Gwyn MacFarlane.

Remembering Bill Lear: "Radio Hits the Road" by Michael Lam, *American Heritage Invention & Technology*, Spring 2000; *Good Morning*, a documentary produced by Ron Blau for The History Channel® with Rick Beyer as executive producer.

Meter Man: Various *New York Times* articles, including "Carl Magee Dead," February 2, 1946; "Meter Matters," *Time*, September 6, 1937.

Diabolical Rays: *Boffin* by Robert Hanbury Brown; "Tells Death Power of Diabolical Rays," *New York Times*, May 21, 1924; various other *New York Times* articles.

Bent Out of Shape: "A Flexible Mind" by Martha Davidson, *American Heritage Invention & Technology*, Winter 2006; "The Straight Truth About the Flexible Drinking Straw," *The Lemelson Center for the Study of Invention and Innovation*, invention.Smithsonian.org.

One for the Road: *Forbes Greatest Technology Stories* by Jeffrey Young; "John Vincent Atanasoff—The Inventor of the First Electronic Digital Computing" by Kiril Boyanov, a paper delivered at the International Conference on Computer Systems and Technologies—CompSysTech'2003.

A Man Called Veeblefetzer: "Al Gross, Inventor of Gizmos with Potential, Dies at 82," *New York Times*, January 2, 2001; author interview with Professor Ted Rappaport, Department of Electrical and Computer Engineering, University of Texas at Austin, July 31, 2008.

The Mystery of the Vanishing Vapor: *They All Laughed* by Ira Flatow; "Making Teflon Stick" by Anne Cooper Funderburg, *American Heritage of Invention & Technology*, Summer 2000.

The Magic Box: *They All Laughed* by Ira Flatow; *Copies in Seconds* by David Owen.

A Real Wonder: "Dr. W. M. Marston, Psychologist, 53," *New York Times*, May 3, 1947;"The Truth about the Lie Detector" by Jack Kelly, *American Heritage Invention & Technology*, Winter 2004.

The Unsinkable Ship: *The Second World War, Volume 5: Closing the Ring* by Winston S. Churchill; "The Floating Island" by Paul Collins, *Cabinet*, issue 7, Summer 2002.

Miracle Poison: "From Poison Gas to Wonder Drug" by Beryl Lieff Benderly, *American Heritage*

Invention & Technology, Summer 2002; *Serendipity: Accidental Discoveries in Science* by Royston M. Roberts.

A Rhino in Normandy: *GI Ingenuity* by James Jay Carafano; *Breakout and Pursuit* by Martin Blumenson; "Eisenhower Hails G.I. for Tank Idea," *New York Times*, June 6, 1964.

Cooking with Radar: *They All Laughed* by Ira Flatow; Raytheon Company archives; interviews with Norm Krim and John Ossepchuk at Raytheon, June 28, 2001.

The Sweet Smell of Success: "Cat Litter" by Curt Wohleber, *American Heritage Invention & Technology*, Summer 2006; "Edward Lowe Dies at 75; A Hunch Led Him to Create Kitty Litter," *New York Times*, October 6, 1995.

The Admiral and the Insect: "Stalking the Elusive Computer Bug" by Peggy Aldrich Kidwell, *IEEE Annals of the History of Computing*, vol. 20, no. 4, 1998; "Amazing Grace" by J. M. Fenster, *American Heritage Invention & Technology*, Fall 1998.

Head Case: *Looking for the Edge*, a documentary produced for The History Channel® by Barbara Moran, with Rick Beyer as executive producer.

Walk the Dog: *Serendipity: Accidental Discoveries in Science* by Royston M. Roberts; *Lemelson-MIT Inventor of the Week Archive*, web.mit.edu/invent/i-archive.html.

Beauty and the Boater: "Marion Donovan, 81, Solver of the Damp Diaper Problem," *New York Times*, November 18, 1998; *Lemelson-MIT Inventor of the Week Archive*, web.mit.edu/invent/i-archive.html.

The Coolest Thing on Ice: *Zamboni: The Coolest Machine on Ice* by Eric Dregi; "The Zamboni Story," www.zamboni.com; *Looking for the Edge*, a documentary produced for The History Channel® by Barbara Moran, with Rick Beyer as executive producer.

Einstein's Brain: *Strange Brains and Genius: The Secret Lives of Eccentric Scientists and Madmen* by Clifford A. Pickover; "Doctor Kept Einstein's Brain in Jar 43 Years" by Bill Toland, *Pittsburgh Post-Gazette*, April 17, 2005; "Einstein's Brain," Physorg.com, January 21, 2005, www.physorg.com/news2778.html.

Surfing Safari: "A Gadget That Taught a Nation to Surf: The TV Remote Control" by Lisa Napoli, *New York Times*, February 11, 1999; "Remembering the Remote Control Inventor" by Randy Frank, *Design News*, April 9, 2007.

Oops!: *Mothers of Invention* by Ethlie Ann Vare and Greg Ptacek; "History" at liquidpaper.com.

Who Invented the Laser?: *They All Laughed* by Ira Flatow; author interview with Ted Maiman, 2001.

Frustrated Fashion Designer: "Russell Colley, Designer of Spacesuits, Is Dead at 97," by Robert

McG. Thomas, *New York Times*, February 19, 1996 (Thomas was one of the great obituary writers of all time. A sampling of his best work can be found in the book *52 McG's*. He died far too young at age sixty); *Smithsonian Annals of Flight Number 8: Wiley Post, His Winne Mae, and the World's First Pressure Suit* by Stanley Mohler and Bobby Johnson.

Heart of a Tigger: "Comic Patents Artificial Heart," *New York Times*, July 20, 1963; "Winchell's Heart," *Time*, March 12, 1973; *Jarvik Heart*, www.jarvikheart.com/basic.asp?id=72.

From Hiss to Bang: *Three Degrees Above Zero* by Jeremy Bernstein; "Penzias and Wilson Discover Cosmic Microwave Radiation," *A Science Odyssey: People and Discoveries*, www.pbs.org/wgbh/aso/databank/entries/dp65co.html.

Patently Absurd II: Based on patents found in the U.S. Patent Office.

The Kernel Quest: "Our Inner Nerd" by Gail Collins, *New York Times*, December 31, 1995; "Orville Redenbacher, Famous for His Popcorn, Is Dead at Age 88," Robert McG. Thomas, *New York Times*, September 20, 1995.

Future Shock: Thanks to YouTube, I was able to watch Doug Engelbart's entire demonstration, videotaped back in 1968, and be a first-person observer as he unwrapped the future. The video is split into nine parts, the first of which can be found at www.youtube.com/watch?v=JfIgzSoTMOs, *Forbes Greatest Technology Stories* by Jeffrey Young; "The Making of the Mouse" by Alex Soo Jung-Kim Pang, *American Heritage Invention & Technology*, Winter 2002.

For the Record: Various articles in the *New York Times*.

Lo and Behold: Author interview with Len Kleinrock, 2001.

Note to Self: E-mail from Ray Tomlinson (no, really!) March 20, 2009; "Merchant@florence Wrote It First 500 Years Ago" by Philip Willan, *Guardian*, July 31, 2000.

Sticking Power: *Serendipity: Accidental Discoveries in Science* by Royston M. Roberts; *Breakthroughs* by P. R. Nayak et al.

Celebrity Patents: Based on patents found at the U.S. Patent Office.

The Happy Non-Hooker: *Looking for the Edge*, a documentary produced for The History Channel® by Barbara Moran, with Rick Beyer as executive producer.

Three Mathematicians: *The Google Story* by David A. Vise.

Space 2.0: "The Sky's the Limit" by Chris Taylor and Kristina Dell, *Time*, November 29, 2004; "The Space Cowboys" by Cathy Booth Thomas, *Time*, February 22, 2007.

ACKNOWLEDGMENTS

An author gets to put his or her name on the cover, but there is always a host of other people whose help is invaluable to making a book a reality. That's as true for this volume, the fourth in the *Greatest Stories* series, as it was for the first. The life of this venture now spans almost a decade, with no end in sight, and I am deeply grateful to all who have traveled down this road with me.

Top of the batting order has to be Artie Scheff, who hired me in 1997 to produce the *Timelab 2000* history minutes for The History Channel®. That paved the way for these books, as well as much of the documentary film work I have done since then.

You changed my life, Artie! I have been very proud to be associated with the network and the many talented people who work there. I would especially like to thank Carrie Trimmer, who has quietly moved mountains year after year to help make this series a success, and Dr. Libby O'Connell, who has always been incredibly supportive.

I have been working for years with the Levine-Greenberg Literary Agency. Arielle Eckstut, my first agent, played a critical role in getting these books off the ground. I am now ably represented by Jim Levine. Jim is the opposite of the agent stereotype you see in movies. He is quiet, thoughtful, and incredibly good at what he does. He and the many other people at Levine-Greenberg are a pleasure to work with.

I was talking with a friend and fellow author recently, who complained that he has gone through five different editors on four books. I have worked with only one editor, the (insert superlative) Mauro Dipreta, and I wouldn't have it any other way. I'm running out of nice things to say about him, but I'll keep trying! Associate Editor Jennifer Schulkind has also been a terrific partner on this book. And I would like to thank the many other folks at HarperCollins, some of whom I know, some of whom I don't, who have worked tirelessly on the series. Let me specifically single out Lisa Sweet, Beth Mellow, and Teresa Brady, without whose publicity efforts you might never have heard of these books.

Help for this particular book came from many sources. The LaPorte Indiana Historical Society sent me lots of great information on undertaker/telephone inventor Almon Strowger. Ditto for the Delaware

County (Pennsylvania) Historical Society, regarding lawyer/matchbook inventor Joshua Pusey. My friend Harry Forsdick was kind enough to put me in touch with his friend Ray Tomlinson, inventor of e-mail. (Perhaps I should thank Ray as well, since without e-mail I don't know how I could ever put these books together!) The descendants of early movie-camera pioneer Louis Le Prince, TV inventor Philo Farnsworth, and Ferris wheel creator George Ferris were all generous with their assistance. And there were many others who fielded my queries with patience and offered much useful information.

I also want to thank Barbara Moran, Ron Blau, Julie Rosenberg, and Kate Raisz, talented documentary producers who all contributed ideas for this book.

The National Archives and the Library of Congress are two invaluable national treasures. Any author looking to illustrate American history with archival photographs, drawings, and maps knows what amazing resources these are. Visiting again in April of 2009 to do research for this book, I found their staff always willing to go above and beyond in offering assistance. I would especially like to mention Patrick Kerwin, manuscript reference librarian at the Library of Congress, who helped me quickly put my hands on photos of Mahlon Loomis. While talking about libraries, I wouldn't want to leave out the Cary Memorial Library in Lexington, Massachusetts, where the librarians are always gracious and helpful.

I am deeply indebted to the many people who managed to take the manuscript and a bunch of pictures and magically create the book you hold in your hands. I'd like to thank Leah Carlson-Stanisic, who developed the interior design and laid out the pages; production manager Jessica Peskay; and Mucca Design for the cover. Special thanks to copyeditor Olga Gardner Galvin, who painstakingly turned a typo-ridden manuscript into something readable, thus helping me keep up the pretense of being literate.

Finally, I would like to heap thanks and praise on my wife and co-conspirator, Marilyn Rea Beyer. The smartest thing I ever did was marry this lovely, intelligent, caring person. She has not only made these books immeasurably better, she is a most amazing partner on this great adventure of life.

Photo Credits

Unless otherwise noted, photo credits for each page are listed top to bottom, and images are listed only the first time they appear. Credits for pages not listed can be found on the facing page. Every effort has been made to correctly attribute all the materials reproduced in this book. If any errors have been made, I will be happy to correct them in future editions.

Abbreviations

LOC: Library of Congress
NARA: National Archives and Record Administration
SI: Smithsonian Institution
MEPL: Mary Evans Picture Library
USPTO: United States Patent Office

Page 2: Author. **Page 3:** LOC. **Page 4:** LOC; Photo by Stan Sherer. **Page 5:** LOC. **Pages 6–7:** University of Iowa Libraries (all except lower right); LOC. **Page 8:** LOC. **Page 9:** Marilyn Shea, University of Maine. **Page 10:** Author. **Page 11:** Bodleian Library, University of Oxford, Shelfmark: MS. Huntington 214, fol. 18v. **Pages 12–13:** LOC, all except photo of stained glass window by Shaun Martin, reproduced by permission of the vicar and churchwardens of Malmesbury Abbey. **Page 14:** LOC. **Page 15:** MEPL;

Photo Researchers. **Pages 16–17:** LOC (all except plan of Solomon's Temple); Babson College. **Page 18:** LOC; Northwind Archive. **Page 19:** Author. **Page 20:** Crane and Co. Inc.; Huntington Library. **Page 21:** Elaine Koretsky. **Pages 22–23:** LOC (all). **Page 24:** David Roth. **Page 25:** MEPL. **Page 26:** © Bettmann/CORBIS. **Page 27:** Science and Society Picture Library. **Pages 28–29** LOC (all except Herschel and his sister); © Bettmann/CORBIS. **Pages 30–31:** Huntington Library (all except Chappe Portrait); Societé d'Histoire de la Poste ed de France Telecom en Alsace. **Page 32:** Author. **Page 33:** LOC. **Pages 34–35:** Getty Images (all except Francis Scott Key); LOC. **Pages 36–37:** LOC (all except Evans portrait); Science and Society Picture Library. **Page 38:** USPTO; author. **Page 39:** Photo Researchers. **Page 40:** Author. **Page 41:** © Bettmann/CORBIS; University of Iowa Medical School. **Pages 42–43:** LOC (except lower right); © Bettmann/CORBIS. **Page 44:** Getty Images; LOC. **Page 45:** Science and Society Picture Library. **Pages 46–47:** LOC (all except FitzRoy); © Hulton-Deutsch Collection/CORBIS. **Page 48:** Bibliothèque nationale de France. **Page 49:** Author; MEPL. **Pages 50–51:** Illustration by David White; Koninklijke Bibliotheek. **Pages 52–53:** Author (all except Goodyear portrait); Historic New England. **Page 54:** The History Place/USPTO; **Page 55:** LOC; illustration by Paul Kiernan, courtesy The History Channel®. **Pages 56–57:** Otis Elevator Company (all). **Pages 58–59** (all). **Page 60:** Academy of Natural Sciences, Ewell Sale Stewart Library and the Albert M. Greenfield Digital Imaging Center for Collections; LOC. **Page 61:** The American Philosophical Society. **Page 62:** Heimatmuseum Gelnhausen; Science and Society Picture Library. **Page 63:** LOC. **Page 64:** LOC; **Page 65:** © Bettmann/CORBIS; Coca-Cola. **Pages 66–67:** LOC (all). **Pages 68–69:** Maine State Museum (all except patent); USPTO. **Page 70:** © CORBIS. **Page 71:** LOC, UISPTO. **Pages 72–73:** Cumberland Packing Corporation (all except microscope); © Tom Grill/CORBIS. **Page 74:** LOC. **Page 75:** Cincinnati Historical Bulletin (patent and text); LOC. **Pages 76–77:** USPTO (all). **Page 78:** Science and Society Picture Library. **Page 79:** USPTO; William Huettel. **Page 80:** USPTO; **Page 81:** LaPorte County Indiana Historical Society. **Page 82:** LOC; USPTO. **Page 83:** D.D. Bean & Sons. **Pages 84–85:** LOC (all except Ferris); Jim Ferris. **Page 86:** Author. **Page 87:** LOC; New York Times. **Page 88:** USPTO, LOC. **Page 89:** Author. **Pages 90–91:** LOC (all). **Pages 92–93:** USPTO (all). **Page 94:** LOC, USPTO. **Page 95:** Hoover Historical Center/Walsh University, North Canton, Ohio; © Bettmann/CORBIS. **Page 96:** ©moodboard/CORBIS. **Page 97:** LOC;USPTO. **Page 99:** Gary Mussell. **Pages 100–101:** LOC (all). **Page 102:** USPTO; Special Collections Research Center, Morris Library, Southern Illinois University Carbondale. **Page 103:** LOC; USPTO. **Page 104:** Author; LOC. **Page 105:** Eli Whitney Museum (Gilbert and artificial heart); Author. **Page 106:** © Bettmann/CORBIS. **Page 107:** Author. **Pages 108–109:** Wisconsin Center for Film and Theater Research (all except patent); USPTO. **Page**

110: USPTO; Jeff Miller. **Page 111:** Johnson & Johnson. **Page 112:** LOC. **Page 113:** USPTO, Intellectual Property Office, United Kingdom. **Pages 112–113:** Photo Researchers (all except Nazi parade); NARA. **Pages 114–115:** Kent Farnsworth (all except patent): USPTO. **Pages 118–119:** LOC (all except Birdseye); Getty Images. **Page 120:** ARS/USDA Scott Bauer; Photo Researchers; copyright Rothamsted Research Ltd. Image provided by Andrew Martin. **Page 122:** Author; USPTO. **Page 123:** LOC. **Page 124:** USPTO. **Page 125:** Getty Images; LOC. **Pages 126–127:** LOC (all). **Page 128:** USPTO. **Page 129:** USPTO; Getty Images; Jim Muntz/Offbeat Trilogy Productions. **Pages 130–131:** USPTO (patents); LOC (cars); POM Parking Meters (Magee). **Page 132:** © Bettmann/CORBIS. **Page 133:** MEPL (all except Watson-Watt); NARA. **Page 134:** Author; USPTO. **Page 135:** USPTO; The Lemelson Center, SI. **Pages 136–137:** University Archives, Iowa State University Library (all except ENIAC); Schoenberg Center for Electronic Text and Image, University of Pennsylvania Library. **Pages 138–139:** Digital Library and Archives, Virginia Tech University (all except Dick Tracy); © Tribune Media Services, Inc. All Rights Reserved. Reprinted with permission. **Page 140:** Dupont. **Page 141:** Photo by Daniel Schwen, commons.wikimedia.org/CC-BY-SA-3.0. **Pages 142–143:** Xerox (all). **Page 144:**"WONDER WOMAN" #7 © 1943 DC Comics. All Rights Reserved. **Page 145:** © Hulton-Deutsch Collection/CORBIS. **Page 146:** NARA; Susan Langley. **Page 147:** Pete Goodreve. **Pages 148–149:** LOC (all). **Pages 150–151:** National Archives (all). **Page 152:** USPTO; **Page 153:** Raytheon Corporation. **Pages 154–155:** Edward Lowe Foundation (all). **Pages 156–157:** LOC (all except logbook); SI. **Pages 158–159:** The Lemelson Center, SI. **Page 160:** USPTO; **Page 161:** Photo Researchers. **Pages 162–163:** The Lemelson Center, SI (all except patent); USPTO. **Pages 164–165:** Courtesy Frank J. Zamboni & Co., Inc. **Pages 166–167:** LOC (all except Harvey); © Bettmann/CORBIS. **Pages 168–169:** Zenith/LG Electronics USA (all except patent); USPTO. **Page 170:** © Fancy/Veer/Corbis; Liquid Paper. **Page 171:** © Bettmann/CORBIS. **Page 172:** AT&T Archives; **Page 173:** © Bettmann/CORBIS; NASA. **Page 174:** NASA. **Page 175:** University of Akron, BF Goodrich Collection; USPTO. **Page 176–177:** USPTO (patent); © Bettmann/CORBIS. **Pages 178–179:** NASA (all). **Pages 180–181:** USPTO (all). **Page 182:** Orville Redenbacher's ® ads used with permission from ConAgra Foods, Inc., Omaha, Nebraska. **Page 183:** Con Agra Ketchum; Purdue University Libraries, Archives, & Special Collections. **Pages 184–185:** Douglas Englebart/Bootstrap Institute. **Page 186:** New York Times; USPTO. **Page 187:** NASA (all except Goddard in cap); LOC. **Pages 188–189:** Len Kleinrock (all). **Page 190:** Mike Ross www .corestore.org. **Page 191:** © Ed Quinn/CORBIS. **Pages 192–193:** 3M. **Pages 194–195:** USPTO (patents); LOC (all others). **Pages 196–197:** Author. **Pages 198–199:** Google (all except Brin and Page). **Pages 200–201:** New Mexico Space Authority (except SpaceShipOne); © Gene Blevins/CORBIS.